2 8

TROPICAL MACROFUNGI

TROPICAL MACROFUNGI
some common species

M. H. ZOBERI, M.Sc., Ph.D. London; F.L.S.

Department of Biological Sciences,
University of Ife, Ile-Ife, Nigeria

MACMILLAN

First published 1972

Published by
THE MACMILLAN PRESS LTD
London and Basingstoke
Associated companies in New York Toronto
Melbourne Dublin Johannesburg and Madras

SBN 333 13887 2

Text set in 11/13pt. Monotype Bembo, printed by letterpress, and bound in
Great Britain at The Pitman Press, Bath

321813

This book is dedicated to my wife
Zohra Zoberi

PREFACE

Most of the tropical libraries are quite inadequate in helping students of mycology. The difficulties experienced in identifying even the commoner and more conspicuous fungi have deterred many a novice, and our knowledge of tropical fungal flora has suffered accordingly. It is hoped that this little book will enable the user to identify some of the attractive fungi which he may encounter. If in so doing a few of my readers are encouraged to take a fuller interest in our fungus flora this book will have been more than justified.

There are so many fungi in the tropics that only a representative selection can be included in a book such as this. Limiting the scope to the more common macrofungi has enabled me to deal in greater detail with some of the more interesting, and often beautiful specimens. However, some rare or occasional species which are exceptionally striking, and especially interesting examples of certain groups, have also been included. In preparing both the descriptions and illustrations, we have employed living fungi, and most of these have been deposited in the University of Ife mycological Herbarium.

The reader should not be dismayed by the scientific names of the fungi; they are preferred to other names because they are precise and internationally understood. I have given some common English names in a number of cases where these seem appropriate. Local names have largely been omitted because they are usually of very little general importance, being often inprecise.

An effort has been made to minimise the use of technical terms. However, in order to guide the amateur naturalist, a glossary is given at the end of the book.

ACKNOWLEDGEMENTS

Persons connected with several institutions were very co-operative in letting me work in the herbaria, using the libraries, supplying information about the species discussed in this book and reading my manuscript. Their names are listed here in appreciation of the favours they granted.

Dr D. M. Dring, (Royal Botanic Gardens, Kew); Professor R. Heim, (Muséum National d'Histoire Naturelle, Paris); Professor C. T. Ingold (University of London); Professor J. A. Nannfeldt (University of Uppsala); Dr D. N. Pegler and D. A. Reid (Royal Botanic Gardens, Kew).

It is also a pleasure to express my thanks to my wife, Mrs Zohra Zoberi, whose illustrations have contributed much to the usefulness of this book. Her skill has portrayed the beauty of form in a manner which cannot adequately be conveyed by a verbal description.

CONTENTS

xii *Contents*

LIST OF ILLUSTRATIONS

LIST OF PLATES

(appearing between pages 64 and 65)

INTRODUCTION

In classical botany, fungi are usually regarded as plants, but this is of doubtful validity. If considered superficially, fungi exhibit some of the features of plants; the mushroom, with cap and gills, which catches the eye of the observer, grows on the ground and appears to be rooted like a plant. The same applies to those fungi which grow on living or dead tree trunks; they are rooted in the bark and some of them possess tubes rather than gills. But if these fungi are examined in microscopic detail, one would no longer hold the view that one is dealing with a plant.

Fungi are similar to green plants in being fixed and possessing cell walls, but they differ fundamentally in their lack of chlorophyll and therefore cannot photosynthesise their own organic food from carbon dioxide and water. They are therefore forced to live as saprophytes on dead organic matter, or as parasites on other living organisms.

The systematic study of fungi is about 250 years old, but the use of this group of organisms in the fermentation of wines and the baking of bread has been known to man for thousands of years, and it is therefore not surprising to find many references to these uses in the writings of the ancient Greeks and Romans. Yet even today, in a science-conscious world, a world in which the surface of the moon has become the playground of the world's scientists, very few people appreciate the importance of fungi.

Fungi play an important role in our everyday life. They are responsible for the decomposition of organic matter, and as such affect us directly by destroying our food, textiles, timber, plastics, leather, rubber, and even glass. They are the cause of the majority of known plant diseases and many of the diseases of animals and men; they are the basis of a number of industrial processes involving fermentation, e.g. Cocoa bean; they are employed in the commercial production of many organic acids, vitamin preparations and a number of very valuable antibiotic drugs. Fungi are both destructive and beneficial to agriculture; on the one hand they damage crops by causing plant diseases, and on the other they increase the fertility of soil by inducing various changes in it. The fruit bodies of some fungi, such as the mushrooms, truffles, morels, and the sclerotia of *Pleurotus tuber regium* are edible. Although their nutritive value is not high, the fruit bodies supply vitamins and are valuable as appetisers. Species must be examined and identified carefully before they are utilised for gastronomic purposes.

Fungi are not solely the concern of the mycologist, but are regarded as very useful tools by researchers in other scientific fields such as cytology, genetics and biochemistry. However, a tendency has developed among certain workers to discount the value of descriptive mycology, and due to the increasing demand of scientists by other disciplines, it seems that before long there will scarcely be anyone capable of identifying even the more common fungi with authority. Yet the importance of the descriptive work based on the field studies will still be of unquestionable value, for until the specimen has been studied in its natural environments, whether it is a fungus or a bacterium, there is little use in studying its behaviour in the laboratory. Only by bringing all the different aspects together is it possible to make good scientific progress. The amateur naturalists, the professional systematists, and the applied biologists all have valuable contributions to make in the disclosure of the secrets of Nature.

The entire fungal kingdom comprises about 150 000 species, whilst many new species are described each year. Although their fruit bodies show a very wide range in size, shape and morphology they all serve the same function, i.e. production and dispersal of the spores resulting from the process of sexual reproduction.

It was the aesthetic appeal of the macrofungi that attracted me during my seven years of study in the tropics, and it is the purpose of this book to introduce these insignificant-looking objects of the hidden world to the reader. Beauty appeals to us through form, colour and smell. It is enhanced by the setting of the object, but once the interest is aroused, we are led to plunge deep into the waters of learning.

The information of the tropical genera and species presented in this book is based on the observations of the fresh specimens and a detailed study of literature, supplemented wherever necessary by the examination of herbarium material collected and classified by other workers.

COLLECTION AND
PRELIMINARY OBSERVATIONS

The macrofungi, such as the mushrooms and toadstools, grow abundantly in the tropics and provide an excellent opportunity for study. Unfortunately in the tropics, fleshy fungi develop and decompose very quickly; it is rather difficult to examine all these specimens before they decay and is certainly impossible to record all their macroscopic and microscopic features, to sketch or paint, and to prepare satisfactory dried or preserved specimens for the herbarium at the same time. Obviously to do all this, specimens should be examined repeatedly and one must plan to study the living fungi over several seasons.

Nearly all amateur field mycologists at first collect specimens which are unhealthy, or they collect very small quantities. It is desirable to collect several healthy specimens at different stages of development, so that one or more of them may be cut across to observe the form of the hymenium, the structure of the stipe and other microscopical details, while yet other fruit bodies may be needed for spore print and for preservation as museum specimens. A young collector should assume that, sooner or later, he will require samples of each collection for sending to the experts for their advice or to other mycologists for exchange purposes.

The collecting technique varies with different groups of fungi; for fleshy fungi, whether collecting for purely scientific or for gastronomic purposes, careful identification is needed. When required for the table, a thorough cleaning, removal of stipe, preferably cutting it off close to the cap, is desirable at the time of picking. For an unidentified specimen, the whole fruit body must be picked intact from the ground or other substratum, special care being taken not to cut or break the stem. Unless the *whole* fruit body is dug up it will be difficult or impossible to know whether the stipe is provided with a volva or cup at the base, whether its base is bulbous or hairy, attached to other plant stems, or forms a pseudorhiza. Wood-inhabiting fungi may have to be cut out and removed with the bark and a portion of the wood upon which it is growing. It is important to identify the hosts and other substrata to which the fungi are attached.

It is a great pleasure for a collector to have in his herbarium the entire range of fungi indigenous to a region and, with the help of co-workers and specialists, to identify as many as possible. Such work adds to our knowledge of distribution, and often provides useful specimens for further study. If one works in a region which already has many records of fungi, one may be able to make useful contributions by re-studying old specimens and eliminating doubtful records.

Since it is difficult to identify, or even collect, all the fungi of a region, it is desirable to specialise in small groups such as the Gasteromycetes, or in the fungi of special habitats, e.g. aquatic or marine, rather than randomly collect all kinds of fungi and spend much time on common, well-known species. However, useful information may be obtained by constant observation of fungi for example the sequence, month by month, season by season and year by year, of fungus production in marked plots, or on fallen logs of known history, etc.

A young collector needs very little equipment; the main essentials are a pocket lens with a magnification of about six times, a pair of scissors, forceps, a good knife, a chisel or a cutlass, jars of preservatives, a few cardboard boxes, envelopes, labels, a pencil; and a note book to enter the details of the macroscopic features, the date, locality and habitat of the specimens collected.

In the laboratory, one needs reasonable working space on tables, and a few shelves or cabinets; it is very important to keep the place as neat as possible. A good microscope, which need not be very expensive, some chemicals, glass slides, mounted needles, a sharp razor and certain other small tools are also needed.

As soon as possible after collection the specimens should be brought into the laboratory for a more detailed inspection. They should be laid out on the table in respective groups and should be examined according to one's interests. Some of the specimens will need immediate examination, others will need pressing, while still others can be left in a preservative or dried for future study. Procedure will vary with individual workers and the kinds of fungi collected, but a suitable scheme should be worked out. The following procedure may be adopted since it seems appropriate to the tropics. The specimens may be collected during the day, preferably during the morning, and all the field data (colour, size, manner of growth, place of growth, texture, smell, taste, etc.) should be entered in the field notebook. In the laboratory, the microscopic observations and preparations of sporeprints and dried specimens should be carried out promptly. It is wise to preserve a suitable specimen in alcohol-formalin. Painting or sketching may be postponed for subsequent seasons.

The microscopic features which should be studied in fresh material are the size, shape and ornamentation of spore and distribution of the pigments; other features, such as the hyphal details and nature of cystidia and basidia can be studied in preserved materials.

MICROSCOPIC EXAMINATION

The importance of microscopic examination of the fungal tissue in clarifying the taxonomy of certain groups of macrofungi has only recently been appreciated, the diagnostic features of fungi being increasingly based on microscopic investigations. It is essential, therefore, to prepare good mounts and keep a record of the observations made. A systematic study of the fungus under various magnifications will enable the worker to understand the morphology of his fungus. Subsequent observations will become easier when a specimen is collected a second time, as it will be checked against the notes of the previous collection and gaps in the knowledge can be filled with new information.

The simplest technique of preparing amount for microscopic study is to remove a minute piece of tissue from the hymenial surface and place it on a glass slide in a drop of 10 per cent potassium hydroxide solution, covering it with a cover-glass. After a few seconds, the cover-glass may be gently tapped to spread out the tissue, so that spores, basidia, cystidia, gloeocystidia, etc. may be easily observed, measured and drawn. Similar squash preparations can be made in Melzer's solution to observe whether the spore walls give a blue amyloid reaction, or a red-brown dextroid reaction.

Squashes can also be prepared either from tissue taken from the upper surface of the pileus to observe the tomentum and pileocystidia, or from the surface tissue of the stipe to observe the caulocystidia.

With many kinds of fungi it is necessary to cut sections to obtain essential data on internal structure. This seldom requires the use of a microtome; beautiful free hand sections can, with practice, be made using a safety razor blade which must be sharp. Small structures, such as the triangular segments from a fresh pileus, can be soaked in water for a few minutes to become turgid, are placed in the cleft end of a piece of pith. Drops of water are placed on the razor blade, and after cutting, the sections are brushed into a watch-glass containing water, then mounted on slides. In this way the relative positions of hymenial elements, the structural details of the hymenium, the arrangement and orientation of the hyphae and the presence or absence of a cuticle, pilocystidia or surface hairs may be observed.

Finally, in order to observe the hyphal structure, thick sections of a fruit body may be cut with a razor blade and soaked in 10 per cent potassium hydroxide solution for 5–10 minutes. The hyphal structure may then be examined easily after teasing apart the tissue with two very fine pointed needles, under a binocular dissecting microscope.

CHEMICAL TESTS

Several chemical tests are used as criteria in the identification of fungi. The following reagents and dyes may be prepared and kept handy to be used when required.

Melzer's Reagent (modified)

Chloral hydrate	100·0 g
Potassium iodide	5·0 g
Iodine	1·5 g
Distilled water	100·0 cm³

The potassium iodide should be dissolved in a small quantity of water first; the iodine should then be mixed in and the mixture thoroughly stirred until dissolved before the remaining water and chloral hydrate are added.

The use of Melzer's Reagent is confined to hyaline or light coloured structures and to the colourless inner wall layers of pigmented spores. The specimen must often first be wetted in ammonia for a few seconds on a glass slide; then the ammonia should be completely removed with filter paper and a large excess of the reagent added.

The reaction, depending upon the final coloration of the preparation, is called amyloid (or pseudoamyloid) if positive, and inamyloid when negative. The amyloid reaction is nearly blue-black, pseudoamyloid is brownish, while the inamyloid reaction is yellow to nearly colourless; individual spores of *Lepiota* stain reddish – brown, a dextrinoid reaction.

Potassium hydroxide (KOH) 10 per cent in distilled water

Used with *Cystoderma*, *Gymnopilus* it darkens certain layers of pileus cuticles in some species, but this reaction is not observed in other species. In *Crinipellis mirabilis*, the epicuticular hairs become grey or green in KOH (Singer, 1942).

Concentrated sulphuric acid (H_2SO_4)

The colour of spores of certain Coprinaceae changes from black to pale blue-grey, while in other species there is no reaction and the black pigment is resistant. A drop of pure acid if applied to the gills of *Amanita phalloides* (a very poisonous mushroom) gives a pinkish-lilac colour.

Ammonia (NH$_4$OH) 50 per cent strong ammonia solution in distilled water. The colour of cystidia in *Stropharia*, *Naematoloma* and *Pholiota* turns a bright yellow. The trama of *Xeromphalina caulicinalis*, and other closely related species, turn red.

Sulpho-vanillin

Chemically pure vanillin	0·5 g
Distilled water	2·0 cm^3
Pure sulphuric acid	4·0 cm^3

The solution is a deep rich yellow colour, and should be filtered through glass wool, handled carefully and must be used on *fresh* material. The reagent is commonly used for *Russula* species. A few drops on the stipe of *R. rosea* turns it a deep red colour. For cystidia, mount a thin piece of cap cuticle or gill in sulpho-vanillin and examine under the microscope, the cystidia are stained in about five minutes.

Guaiac (ordinary guaiac tincture)

The oxidases present in fungi react with the guaiaconic acid. A blue (or green) to purple spot on the surface of the stipe indicates a positive reaction.

Guaiacol (solution in water)

Used in Russulaceae, Tricholomataceae, and Amanitaceae. A salmon colour-orange to rose or bluish pink spot on the base of the stipe indicates a positive reaction.

Phenol (carbolic acid): 2 per cent in distilled water

Used with Russulaceae, Amanitaceae and Tricholomataceae; a chocolate or deep purplish violet colour indicates a positive reaction. If, after 20 minutes, there is no distinct change the reaction is regarded as negative.

Ferrous sulphate (FeSO$_4$): 10 per cent in distilled water

Used with *Russula*, *Tricholoma*, *Tricholomopsis*, and many Boletaceae. The colour reactions are of several categories: (1) negative reaction – no colour change; (2) some shade of olive, green, blue-green or blackish-green colour of the context or surface of the stipe; (3) all shades from pure pink to grey; (4) blue or green-blue to slate grey; (5) a variable colour effect on the cuticle of the pileus.

Henry's reagent

Thallium oxide	2 g
Concentrated nitric acid	1 cm^3
Concentrated hydrochloric acid	4 cm^3
Sodium bicarbonate	1 g
Distilled water	10 cm^3

This is a poisonous reagent. Useful for identification of *Agaricus xanthodermus* from other edible species; a few drops applied to the cuticle of the cap gives a brick-red positive reaction, on *A. xanthodermus* there is no such reaction.

Schaeffer's cross reaction (aniline + HNO$_3$)

A streak of concentrated HNO$_3$ is made on the pileus of *Agaricus* specimens, then a crosswise streak is made with aniline oil. The positive reaction is an orange-red to fire-red discoloration.

Aniline (aniline oil and aniline water)

Pure aniline oil, or 50 per cent in distilled water. It gives a red to copper-red coloration on the context of the stipe and causes a central stained spot, surrounded by a grey or brightly coloured zone, on the gills of *Russula*, *Boletus* and *Agaricus* species, and for several genera in the Aphyllophorales.

Phenol-aniline: 2 per cent in distilled water

A few drops of aniline oil with phenol in water. Must be applied on dried materials. The reaction is from nil to nearly black after prolonged exposure.

Alpha-naphthol

The reagent need not be accurately mixed; a few crystals are dissolved in about 2 cm^3 of 90 per cent alcohol and then about 4 cm^3 of water are added. The solution reacts rapidly with the context of the stipe of *Marasmius grandisetulosus* causing a deep wine coloration.

Pyrogallol

5 per cent in distilled water is said to give richly coloured reactions with the context of *Russula*, *Marasmius* and other mushrooms.

Pyramidon (saturated solution in distilled water)

A positive reaction indicates a light lilac colour on the context of the stipe; used in *Tricholoma* and *Russula*.

Formaldehyde 40 per cent in distilled water

A positive reaction indicated by a change in the colour of the context. It is a slow reaction, 20 minutes being required to accomplish the change. Used in *Tricholoma*, *Russula* and *Boletus*.

Aceto-carmine (for carminophilous basidia)

Test must be made on dried material. Pour 2–3 drops of the reagent on a small piece of gill on a glass slide, heat gently without allowing the material to dry out. Finally mount in fresh aceto-carmine, apply cover-glass and tap it with a pencil to splay out the material; examine under an oil immersion lens. Basidia with black staining granules indicate a positive reaction.

A SYNOPSIS OF THE ORDERS AND FAMILIES INCLUDED IN THIS BOOK

The word macrofungi is a general term used to cover a multitude of higher fungi having large, macroscopic, complex fruit bodies. The group includes mainly terrestrial species of diverse form and habitat, and consists of Ascomycetes, in which the spores are produced inside a globose, clavate, cylindrical or occasionally filiform sac-like structure known as an ascus; and Basidiomycetes in which the spores are borne externally on the surface of special structures called basidia. A nuclear fusion occurs during the development of both asci and basidia.

ASCOMYCETES

The Ascomycetes according to Martin (1961) are subdivided into Hemiascomycetidae and Euascomycetidae, but in this book we are concerned only with Euascomycetidae which may conveniently be subdivided into four well-defined series: the Plectomycetes, the Pyrenomycetes, the Discomycetes and the Laboulbeniomycetes. We shall briefly describe two of these.

Pyrenomycetes. In this series club-shaped or cylindrical asci, which are usually persistent, develop in basal tufts or in a layer which lines the entire inner surface of the perithecium. The perithecium may or may not be embedded in a fruit body. The series Pyrenomycetes is divided into about nine or more orders. The following brief survey of Sphaeriales and Hypocreales follows Martin's Key (1961).

Sphaeriales

Numbering some 4 500 species this order is very common on wood, bark, fallen or living leaves, twigs and branches of trees, plant compost and dung. This order includes all those fungi possessing dark, coriaceous or carbonaceous, spherical or pear-shaped, ostiolate perithecia in which unitunicate asci along with paraphyses, if present, form a persistent hymenial layer or a basal tuft. The perithecia either develop directly from the loose mycelia or associate with a well-developed fruit body resting on the surface or sunken within its context. The asci are generally clavate, inoperculate, and may, with paraphyses and periphyses, form a hymenium. The ascospores are violently discharged from the asci and vary greatly in their form. At this stage of our knowledge the classification of the Sphaeriales is confused, no two students of these fungi agreeing on the limits or the chief characters of the group. We are concerned in this book with only two families: Sordariaceae, with superficial perithecia without an easily discernible fruit body; and Xylariaceae with a stroma entirely composed of fungal elements.

Hypocreales

This order contains about 800 species, including a number of forms which live saprophytically on green plants, fungi, insects, and other organic matter. A few of the parasitic species grow on fruit bodies of various Basidiomycetes. Generally all the fungi included in this order either possess a brightly coloured ostiolate perithecium with relatively soft or waxy walls, or brightly coloured, more or less soft and fleshy fruit bodies bearing perithecia either on the surface or immersed in the tissues of the fruit body. The order is often subdivided into two large families; the Clavicipitaceae, with long cylindrical asci containing a thick end operculum-like tip; and the Nectriaceae with elliptical to cylindrical inoperculate asci and various-shaped ascospores.

Discomycetes. These fungi bear their club-shaped or cylindrical asci arranged in a hymenium on an open cup-shaped fruit body (apothecium). In one order the fruit bodies develop in the sub-surface soil and in most species remain closed. The epigean forms are subdivided into *inoperculate*, in which the asci release their spores through an apical slit; and *operculate*, in which the ascus tip has a lid-like structure, the operculum, which opens to permit the escape of the ascospores. The spores from the asci are puffed out in clouds, occasionally with a hissing sound. The series Discomycetes contains about four or more orders.

Pezizales

This order is included in operculate Discomycetes and consists of a wide variety of fungi, most of them saprobic, living on dead wood, soil or humus; except those which are edible, these fungi are of little direct importance to us. The members of this order form the large, conspicuous, often brilliantly coloured cups which are encountered growing in the woods during the rainy season. They also contain some of the smallest fruit bodies and often go unnoticed by anyone who is not searching particularly for them. The classification of Pezizales has not been stabilised as yet; we shall discuss only two families here:

Sarcoscyphaceae, which is mainly tropical but possesses some cosmopolitan species and is characterised by its sub-operculate ascus having a thick apical ring, and a plug, or operculum; and Pezizaceae, with operculate asci and cup-, disc-, or lentil-shaped fruit bodies which may or may not be differentiated into stem and cup and are bright-coloured to dark brown, smooth, hairy, or bristly. Many species grow on the ground, others appear on the dung of herbivores, and a number inhabit wood.

BASIDIOMYCETES

The Basidiomycetes, like Ascomycetes, are divided into two sub-classes: the Heterobasidiomycetidae, in which the basidium is either septate or deeply divided and the basidiospores are capable of germinating by repetition. These include the rusts, the smuts and the jelly fungi. And the Homobasidiomycetidae, in which the basidium is neither septate nor deeply divided and the basidiospores usually germinate by a germ tube. The fungi included in this sub-class are very common and are generally known as mushrooms, toadstools, puffballs and stinkhorns. The so-called shelf or bracket fungi (or bird's-nest fungi) are also included in this group.

Heterobasidiomycetidae. Various schemes have been proposed for sub-dividing the Heterobasidiomycetedae into orders and families. Three order names (Martin, 1961) in common use are Uredinales, Ustilaginales and Tremellales.

Tremellales

This order, also known as the jelly fungi, has usually well-developed fruit bodies, often having a jelly-like consistency. However, many species produce waxy or cartilaginous fruit bodies. The Tremellales is common in tropical countries, most of the species being saprophytes, commonly on wood; a few organisms, included in the family Septobasidiaceae, live in association with scale insects, others are destructive parasites while some are beneficial and show promise for use in the commercial production of carotenoids. According to Martin (1961) the order Tramellales is subdivided into nine families; species belonging to only three of these families have been described in this book.

Dacrymycetaceae: having bright yellow or orange, jelly-like or waxy fruit bodies growing during the rainy season on the branches or trunks of living or dead trees. The chief characteristic which distinguishes these fungi is the typical spore-producing apparatus. It consists of an elongated hypobasidium produced on the apex of a binucleate hypha with two long arms, the epibasidia, each of which terminates as a sterigma on which a basidiospore develops.

Tremellaceae with crust-like to stalked fruit bodies; often the entire fruit body is a thin layer of gelatinous hyphae which produces the basidia. Some species are cushion-shaped with a highly wrinkled surface and of a grey, purplish, brown or white colour.

Auriculariaceae is rather a large family, fruit bodies vary from a simple weft of hyphae to the well developed large fruit body of *Auricularia* which is gelatinous and somewhat waxy. The distinguishing feature of this family is the transversely septate basidlum. Most of these fungi are saprobic, others are parasitic on mosses or on economic plants.

Homobasidiomycetidae. The Homobasidiomycetidae are divided into two distinct more or less natural groups. These are Hymenomycetes and Gasteromycetes, and like Pyrenomycetes and Discomycetes of the Ascomycetes are much used as semipopular designations. These may be termed 'series' since these are not always recognised in current taxonomic terminology.

Hymenomycetes. The Hymenomycetes are characterised by basidia which develop in a well defined hymenium which becomes exposed even when the spores are still undeveloped. The series is generally considered to include, according to Martin (1961), two orders: Polyporales (or Aphyllophorales) and Agaricales.

Aphyllophorales

These develop their hymenial layer in various ways on definite, gymnocarpous, papery, leathery or woody, but not soft and putrescent fruit bodies. The hymenium is either unilateral, bilateral or amphigenous. It may be smooth, ridged, warted or spiny, or it may develop inside the tubes or on the surface of the gills. The order is a large one and the system of classification is not very stable. Martin (1961) has recognised six families, we are concerned with four of these and shall discuss them rather briefly.

Thelephoraceae. It is characterised by a more or less smooth, flat and unilateral hymenium which is often resupinate; it seldom takes the form of gills, tubes or spines. Cystidia, or setae, are often present. The spores are non-septate and often very thin-walled, smooth or rough, and hyaline to sometimes coloured. The fruit bodies are often very thin, consisting of only a thin layer of hyphae with basidia resembling a coating of paint on the surface of the substratum. However, the more advanced members have definite, pileate, sessile or stipitate

basidiocarps of papery, leathery or woody texture. Some species are of economic importance and attack a number of important hosts such as oak and grape
vines; others are saprophytes on wood. For many years the Thelephoraceae has
been studied extensively and about 800 species and 35 genera are included in
this family.

Clavariaceae. Some of the most beautiful fungi are included in this family.
Some of these are commonly known as coral fungi because of their intricate
branching system which resembles coralloid growth. Generally, the Clavariaceae are distinguished mainly by the appearance of their fruit bodies, particularly the presence and absence of branching, hymenial characteristics, nature of
spores, and certain other characters such as colour, odour, taste, size and the
manner of growth which can only be observed in fresh material. Now, however,
Corner (1950) has emphasised the importance of hyphal structure and other
microscopic features in distinguishing the genera of this family. In a very
valuable contribution towards clavarioid systematics, emphasising hyphal
analysis, Corner comes to the conclusion that the name Clavariaceae is unde-
sirable because many of the clavarioid genera have no close affinities with each
other and so it is not possible to define the limits of true Clavariaceae in the
taxonomic context of the type-genus *Clavaria.*

Hydnaceae. This family includes pileate or resupinate species. The fruit bodies
possess hymenial teeth on the lower surfaces and may resemble crusts, mushrooms, or corals, but generally the coral-shape fruit bodies are softer and more
gelatinous than those of the Clavariaceae. Some of the fungi included here
are intermediate between Hydnaceae and Thelephoraceae.

Polyporaceae. This family has had much attention and been studied extensively
by many authors. The old-established family, based on the configuration of the
hymenial surface, is being broken down and new families are being established
as the studies on the analysis of hyphal system (Corner 1932) pigmentation,
clamp-connections, hymenial structures, and spores reveal new relationships,
previously unsuspected. Singer (1962) has included Polyporaceae with the
order Agaricales and grouped together all the agaricoid genera, which are
either strictly close to the genus *Polyporus* or show affinities with *Pleurotus,*
Panus or *Lentinus.* Other poroid genera which do not show such a relationship
are retained in the Aphyllophorales. The following are the distinguishing
characteristics of the family: fruit bodies tough, fleshy or leathery; stipe central,
eccentric, lateral, or absent, sometimes a false stipe is present. Hymenophore
weakly to strongly developed if present, it is poroid to alveolar, venose or
lamellate; tubes and gills never free; hymenophoral trama, where present,
distinctly regular to irregular; trama inamyloid with numerous thick-walled
hyphae, occasionally gelatinised; cuticle generally dense; cystidia, if present,

are of metuloid type or leptocystidial; branched setae present in some pore-bearing species, in some species there are hyphal pegs breaking through the hymenium. Usually growing on dead or living wood.

Agaricales

The Agaricales can roughly be characterised as the order consisting of Agarics and Boletes, and includes most species which produce fruit bodies commonly known as mushrooms or toadstools. The fruit bodies are macroscopic, hymenium well-developed; basidia simple, two or four-spored, with apical sterigmata. In modern taxonomy this definition has become too complicated to be expressed in a customary short diagnosis. Singer (1962) considers that in order to identify the species of Agaricales with much more accuracy, microscopic and macroscopic examination should be made of the veil, spores, spore print, mycelium, carpophoroids, stilboids, conditional carpophores, bulbillosis, primordial development, context-structure, hymenial and cortical layers of the fruit bodies, and the sterile tissue of the hymenophore. Chemical, physical and cytological characters should also be taken into account. Based on these criteria, he divides the order into sixteen families, species belonging to nine of these are briefly discussed in this book.

Hygrophoraceae. Are characterised by the viscid cap and gilled hymenophore; gills waxy and thick; basidia 5–7 times longer than spores, two- or four-spored; spores hyaline and thin-walled, small and globose to cylindrical, smooth, inamyloid rarely amyloid; veil absent; hymenophoral trama intermixed, irregular; hyphae inamyloid with numerous clamp connections; found growing on the ground in woods, rarely on decayed wood.

Tricholomataceae. Are distinguished by fruit bodies with homoiomerous flesh and an occasional separation zone between the pileus and the stipe; stipe fleshy, cartilaginous, horny or chordaceous; trama rarely pseudoamyloid; hymenophoral trama never inverse; spore print cream to white; spores uninucleate, inamyloid, amyloid or rarely pseudoamyloid; basidia with one-, two-, three-, or four-spored. On the ground in wood and on living or decaying substrata.

Amanitaceae. Are characterised by their pleutioid, gilled fruit bodies; gills free or subfree; lamellulae numerous, thin, spore print white, pink or greenish; spores usually thin-walled, amyloid or non-amyloid; basidia normal, four-spored; cheilocystidia usually present; hymenophoral trama bilateral; stipe central, often veiled with a volva. Found on the ground in woods, on termite nests or on wood.

Agaricaceae. Are commonly characterised by the furfuraceous to scaly surface of the cap and the presence of an annular veil; pileus often umbonate; hymenophore lamellate; lamellae thin, free; hymenophoral trama regular to irregular;

basidia normal, generally four-spored; spore-print colour variable, hyaline to dark coloured; spores thick-walled, usually binucleate, amyloid to nonamyloid; stipe central; volva often well developed, sometimes rudimentary or absent; context fleshy consisting of amyloid or non-amyloid clamped or unclamped hyphae. Found on ground, in lawns or on dead and living plant tissues.

Coprinaceae. Possess lamellate hymenophores, lamellae with parallel to sub-parallel sides of wedge-shaped, aequihymeniferous or inaequihymeniferous; in the aequihymeniferous genera with wedge-shaped gills the surface of pileus is characteristically cellular; the inaequihymeniferous gills with parallel or subparallel sides usually tend to deliquesce and eventually the whole pileus becomes liquefied and forms drops of spore-coloured liquid. Spores usually dark coloured, with distinctly double or more complex wall, smooth, echinate, reticulate or otherwise ornamented, of various shapes and sizes; basidia normal; lamellae free to subdecurrent; hymenophoral trama regular becoming sub-regular when old; stipe central, with or without veil; context somewhat dry when matured, usually fleshy or membranous, often very fragile, hyphae usually with clamp connections. Found on various substrata such as dung, humus and living or dead plant tissue.

Bolbitaceae. Are recognised by their gilled hymenophores and the epicutis which consists of pyriform or globose, erect cells. The cap is usually hygrophanus; cheilocystidia present; spore print brown, often rust coloured; basidia normally four-spored; stipe central, fleshy to fragile; hyphae in most species are clamped. Found on wood, dung and other substrata, frequently grows on lawns, in gardens and on the ground in woods.

Cortinariaceae. The pileus surface is either a trichodermium or a cutis; hymeno-phore lamellate; hymenophoral trama definitely regular; spore print some shade of brown; spores with a compound wall, germ-pore absent; basidia normal; cystidia often present; stipe central, rarely eccentric, lateral or absent; hyphae generally with clamp connections. Found on the ground in woods, on decayed wood, on rhizomes of orchids and ferns and on palm leaves.

Boletaceae. Pileus scaly, fibrillose, mealy, tomentose, granulose, velutinous, or glabrous, viscoid or dry; hymenophore tubulose, tubes usually easily separable from the context of the pileus; spore print bright-coloured; spores smooth, with homogeneous walls, usually without germ pore; hymenophoral trama bilateral; stipe cylindrical, attenuate, or thickened towards the base, or ventricose to bulbous, solid to hollow, veil often present. Generally on the ground in woods, majority symbiotic with forest trees forming mycorrhiza.

Gasteromycetes. Due to great diversity among the genera of this un-natural taxon, Ingold (1965) considered the gasteromyceteous fungi as being at a highly experimental stage in their evolution. These fungi differ from the Hymenomycetes in having their poorly defined hymenia closed, at least at an early stage of the development of their fruit bodies; their basidiospores are not violently discharged from their sterigmata.

Unable to produce ballistospores these fungi have adopted other means of dispersal; at least one factor is to their advantage, since the ballistospore discharge depends upon fairly high humidity (Zoberi 1964) the hymenium must therefore remain under humid conditions throughout the period of spore discharge. The Gasteromycetes not having this limitation, have adapted to fruiting under dry conditions, which has resulted in diversification of their habit and habitat, or according to Ingold, 'Nature has tried again'.

There are many distinct and fairly convincing suggestions which link the gasteromyceteous genera with others amongst the Hymenomycetes, but all these affinities occur within the Hymenogastraks, only one of the five orders of Gasteromycetes. Thus Ingold (1965) has suggested this order be shifted from Gasteromycetes to Hymenomycetes with a necessary redefinition of the group.

According to Ainsworth (1961) the Gasteromycetes are divided into five orders, we shall discuss four below.

Phallales

The order is characterised by the absence of capillitium in the mucilaginous gleba which often has a putrefactory smell, unpleasant to man, but attractive to certain insects. Usually the gleba develops on a delicate and complex receptacle. The young fruit body, or 'egg', enclosed by a peridium develops at or under the surface of the soil, the peridial layer breaks open through the emergence of the receptacle from the fruit body. Fisher (1933) has divided Phallales into the following two families.

Phallaceae. The species of this family are recognised by their delicate, short lived, usually spongy receptacles developed inside a roundish 'egg'. At maturity the egg splits open and the elongated, hollow, chambered receptacle emerges bearing the gleba either directly on the stipe near the apex, or in most genera, on a more or less campanulate cap which fits over the apex of the stipe. In some species a conical network (indusium) develops from near the apex of the stipe and hangs down around the stipe. The ruptured peridium remains attached to the base of the receptacle and is known as the volva.

Clathraceae. Although this family is closely related to the Phallaceae, it differs in many respects. The gelatinous 'inner peridium' of the egg is divided into segments. The fruit bodies are rather complicated structures. The receptacle is sessile, or stalked, and consists of either network or of variously united arms or branches. The gleba develops on the inner sides of the arms except in the genus

2

Kalchbrennera in which the glebal mass is produced on the outside of the network. The shape of the receptacle is used as a criterion in distinguishing the genera in this family.

Nidulariales

This group of Gasteromycetes is characterised by the nest-shaped, or funnel-like receptacle, which contains one to several hard eggs or peridioles, which are sometimes attached to the wall of the receptacle by means of complicated thread-like structures known as funiculi. The top of the fruit body, in its young stage, is usually closed. The dispersal mechanisms of the following two families show an extraordinary specialisation.

Nidulariaceae. The members of this family are commonly known as 'birds nest fungi' characterised by the small nest-shaped receptacle containing several hard, ovoid peridioles which are attached to the wall of the receptacle by means of a thin cord, the funiculus. Spores enclosed inside the peridiole are generally large, thick-walled and smooth-surfaced.

Sphaerobolaceae. A family containing only a single genus, it is very closely related to the Nidulariaceae, differing only in the mechanism of peridiole discharge. The cup-shaped fruit body contains only a single peridiole which is discharged violently from the inner peridium which suddenly everts.

Lycoperdales

This order is characterised by the abundant capillitium. The glebal tissue and the basidia undergo autodigestion when mature, leaving a mass of dry spores and capillitial threads. The spores are gradually released through a pore or by the breakdown of the peridium. According to Ainsworth (1961) the order is sub-divided into three families, two of which are discussed below.

Lycoperdaceae. Fruit bodies more or less spherical, sessile or with a pseudo-stipe. Peridium divided into two layers, the exo- and endoperidium. Gleba powdery at maturity, consisting of spherical or ellipsoid, usually ornamented spores and well developed capillitium.

Geastraceae. In this family the peridium usually consists of four layers; the endoperidium remains intact and opens through one or more definite ostioles; the three-layered exoperidium splits in stellate fashion, often the outermost layer of the exoperidium separates and remains in the ground, the two remaining layers then split and elevate the intact endoperidium with the glebal mass.

Hymenogastrales

This order includes some species which are intermediate in structure between the Hymenomycetes and the Gasteromycetes. The fruit bodies are commonly subterranean or come above the ground's surface only when they are mature. The gleba is generally fleshy or wax-like, but powdery in *Podaxis*, and slimy. Some species of other genera exhibit a strong odour. On the basis of certain histological similarities the order has a fairly convincing link with the fleshy Hymenomycetes. We are here concerned with only one family.

Podaxaceae. This is a small family closely related to the Secotiaceae. It consists of only a single genus, *Podaxis*. Easily recognised by its resemblance to agarics, it has a stipe and an apical pileus which remains almost close at maturity. The tramal plates, inside the peridium or pileus, break down before maturity to be replaced by the capillitial threads. The powdery glebal mass resembles that of Lycoperdaceae, but the spores resemble those of Secotiaceae, which are brown coloured, double-walled, apiculate with a germ pore.

KEY TO THE FAMILIES DISCUSSED

SPHAERIALES

Perithecia stromatic; stroma fleshy to tough, persistent, entirely of fungal elements; ascospores brown to black.   (*Xylariaceae*)

PEZIZALES

Fruit bodies develop on woody substrata; apothecia often large, stalked, fibrous, ligneous; asci suboperculate not blued at the tip by iodine, narrow, not protruding beyond the general level of the hymenium at maturity.

(*Sarcoscyphaceae*)

Fruit bodies usually on soil or humus; apothecia sessile or stipitate; asci operculate, often blued at tip with iodine.   (*Pezizaceae*)

TREMELLALES

Epibasidia cylindrical, or inflated at the tip, sometimes absent; basidia not septate; probasidia subcylindrical, then narrowly clavate, becoming furcate; spores usually septate, producing conidia.   (*Dacrymycetaceae*)

Epibasidia cylindrical, or inflated at the tip; probasidia subglobose or pyriform, rarely fusiform, not becoming furcate, longitudinally or obliquely septate into 2–4 cells; spores apiculate, asymmetrical.   (*Tremellaceae*)

Epibasidia cylindrical or inflated; probasidia clavate, cylindrical or occasionally ovate, not becoming furcate; septa transverse, usually three; spores cylindrical to allantoid.   (*Auriculariaceae*)

POLYPORALES (APHYLLOPHORALES)

Hymenium smooth, roughened or corrugated; basidiocarp covered with delicate fibrils, membranous, leathery or hard, resupinate to pileate and stipitate.

(*Thelephoraceae*)

Hymenium smooth, radiately ribbed or verrucose; basidiocarp spatulate, infundibuliform or marismatoid, stipitate; basidia clavate, four-spored; spores smooth, hyaline, non-amyloid. (*Podoscyphaceae*)

Hymenium amphigenous, smooth or wrinkled; basidiocarp pileate, clavate or coral-shaped. (*Clavariaceae*)

Hymenium spiny or poroid, rarely lamellate, lining interior of tubes or pits; pores deep or shallow; if shallow sterile on ridges; texture sub-fleshy or leathery or hard, rarely tough gelatinous, but never soft and putrescent.

(*Polyporaceae*)

AGARICALES

Hymenium lamellate, lamellae often rather widely spaced, thick and waxy; basidiocarp pileate; pileus often viscid; pileal trama and stipe homiomerous; hymenial trama bilateral; spore print white; spores smooth, inamyloid.

(*Hygrophoraceae*)

Hymenium lamellate, lamellae not free; pileal trama pseudoamyloid; spores uninucleate; cream to white spore print. (*Tricholomataceae*)

Hymenium lamellate; lamellae free or almost so; spores often amyloid; white spore print; veil often well developed; often produces mycorrhizae (ectotroph); or grows in termite nests. (*Amanitaceae*)

Hemenium lamellate, lamellae free; stipe central, annulus generally present and well developed; spore print purple-brown to sepia; spores pseudoamyloid, surface or trama of basidiocarp often showing characteristic colour reactions with KOH and strong acids. (*Agaricaceae*)

Hymenium lamellate, lamellae usually tend to deliquesce; spore print black, purple-brown, fuliginous, sepia, fuscous, or rarely dull brownish purple.

(*Coprinaceae*)

Hymenium lamellate, epicutis consisting of a continuous layer of piriform or globose cells; pileus hygrophanous; spore print bright rust colour.

(*Bolbitaceae*)

Hymenium lamellate; hymenophoral trama regular; spore print brownish, argillaceous to bright and rich ferruginous—fulvus greyish-olive (rarely) to almost black.   (*Cortinariaceae*)

Hymenium poroid, rarely lamellate; tubes short to long, usually very easily separable from the context of the pileus; spore print pale to deep olive, cinnamon, fawn or lemon yellow.   (*Boletaceae*)

PHALLALES

Basidiocarp sessile or stipitate, expanding into a complex receptacle of a spherical network, or of several appendages united at the top, or spreading; gleba becoming mucilaginous at maturity, usually borne on the inside of the receptacle; peridial sutures present.   (*Clathraceae*)

Basidiocarp a single unbranched elongated receptacle, spherical network never present; gleba becoming mucilaginous at maturity, usually borne on the apex of the receptacle; peridial sutures absent.   (*Phallaceae*)

NIDULARIALES

Basidiocarp nest-shaped; gleba organised into seed-like peridioles which are usually numerous, attached to receptacle by elastic funiculi or embedded in mucilage, not violently discharged.   (*Nidulariaceae*)

Basidiocarp with a single peridiole, discharging violently at maturity; receptacle globose when young, about 2 mm diameter, opening when mature.   (*Sphaerobolaceae*)

LYCOPERDALES

Basidiocarp sessile or with a sterile base formed of sponge-like tissue; capillitium abundant.   (*Lycoperdaceae*)

HYMENOGASTRALES

Basidiocarp with a cap and a true woody stipe, prolonging into the gleba as a percurrent columella; capillitium abundant.   (*Podaxaceae*)

XYLARIACEAE

XYLARIA Hill ex Greville

The species belonging to the genus Xylaria are common throughout the world and are easy to preserve. They usually grow in wet and shady places, either on dead wood or bark, or as parasites on woody plants. All the species vary somewhat in shape and size. However, the general habit is fairly characteristic. The stromata are upright, simple or branched and sometimes forked, more or less club-shaped, cylindrical or fusiform, flesh white or pale buff, black in one species, (*X. dealbata*). In most large tropical species the stroma becomes hollow at maturity; apex apiculate, free from perithecia, often covered with light coloured conidia; the stromatic surface is covered either by a thin fibrous layer or a hard crust varying in colour, splitting in a characteristic manner to expose the perithecial ostioles. Perithecia are usually embedded in the stromatal tissue beneath the crust, but in a few species they are superficial. Spores are uniseriate, non-septate, more or less flattened, with one edge much straighter than the other, dark brown or nearly black at maturity, sharply apiculate, occasionally with hyaline appendages.

Distribution: cosmopolitan.

Practical importance: cause serious diseases of economic plants, among a wide range of hosts. Many cause black lines in timber due to zones of sclerotium-like growth.

XYLARIA POLYMORPHA (Pers. ex Fr.) Grev.

A common cosmopolitan species, usually encountered growing in small clusters on dead stumps of wood, mostly exceptionally polymorphic. The fruit bodies grow towards a source of light and in the laboratory can be made to twist and turn at will. Abnormal fruit bodies, due to insufficient light, are sometimes found growing on the undersides of logs, in hollow trunks of trees and under masses of fallen leaves. A normal fruit body is more or less club-shaped, tapering below, solid, then becoming hollow; 8 cm long and 2·5 cm wide, but in tropical species it is seldom more than 4 cm long and 1·5 cm wide; tips rounded, occasionally obscurely lobed; stalk usually clearly defined; crust brown to black, corky, varying from almost smooth with a tomentose outer layer;

splitting into small rectangular scales and regular network of deep cracks, to slightly mammilate, with minute, closely spaced, obtuse, black warts. Flesh white to pale buff, hard, solid. Perithecia embedded in the flesh; tips of the perithecial necks, with lightly coloured prominent ostioles, protruding from the crust; ostiolar papillae black, slightly convex; asci cylindrical, 200 × 10 μ, containing eight ascospores. Spores black, fusiform, uniseriate, 20–30 × 6–12 μ.

Distribution: cosmopolitan, mostly in Europe, N. America and Asia; widespread in French Guinea, Brazil, Venezuela, Columbia, Costa Rica, Jamaica, Cuba and Africa.

Practical importance: causes serious disease of economic plants, especially beech.

XYLARIA HYPOXYLON (L. ex Fr.) Greville (Fig. 1.)

This species, due to the peculiar shape of its fruit body, is commonly known as the 'candle-snuff' or 'stag's horn' fungus. It is one of the commonest species of *Xylaria* and is usually encountered growing on rotting wood or on the bark of dead trees. Stroma upright, corky, flattened, simple or branched, up to

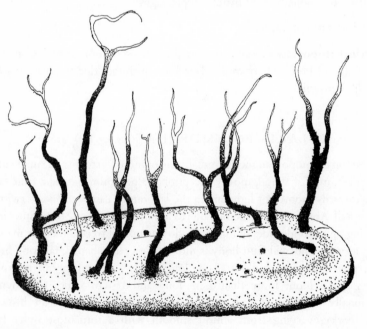

Fig. 1

8 cm in length. The bifurcated tips of the stromata are covered with white-coloured conidiophores which produce vast numbers of white conidia. The conidia can be seen dispersing as a visible cloud when the branches are suddenly jerked. The lower part of the stromata is generally black; perithecia develop beneath the surface showing the protruding papillae of the perithecial necks. Perithecia with conspicuous ostioles; asci cylindrical, 100 × 5 μ, each containing eight spores; spores smooth, black, non-septate, more or less bean shaped, uniseriate, 11–14 × 5–6 μ. Context of the stroma white.

Distribution: cosmopolitan.

Practical importance: grows on a wide range of hosts, causing root rot of apple and other economic plants; also causes black lines in timber which due to zones of sclerotium-like growth.

XYLARIA MULTIPLEX (Kze) Fries

A common *Xylaria* of the tropics, this species is encountered growing on the stumps of wood in forests, in coffee plantations and under bamboo trees. Stromata densely caespitose, slender, occasionally grows to a length of 12 cm but generally very much shorter, 2·5 mm wide, usually pointed and sterile at the tips, tapering below towards a narrow base; strap-like, finely tomentose stalk, which is often once or twice forked; stromata sometimes distinctly nodulose with the perithecia; crust smooth, black, finely divided by longitudinally anastomosing cracks, at first covered by a thin, paler, matt surface layer which splits into narrow subpersistent strips; perithecial ostiole punctate, scarcely papillate; asci 80–90 × 5 μ; spores 9–12 × 3·5–4·5 μ.

Distribution: tropical.

DALDINIA Ces. & de Notaris

A very common genus usually found growing on dead trunks and logs of wood. The fruit body bursts through the surface of the substratum, and is stipitate to hemispherical, or effused and sessile, rubiginous or of various shades of purple; context dark, fibrous, showing distinct concentric zones; perithecia appearing in a single layer just below the outer crust, oval to compressed, ostiole umblicate; asci cylindrical, eight-spored; conidial layer borne on the outside of the young fruit body; conidia hyaline, lightly coloured, minute.

Distribution: cosmopolitan.

DALDINIA CONCENTRICA (Bolt. ex Fr.) Ces. & De Notaris

A very common fungus usually growing on dead trunks and logs of wood. Stroma is hemispherical with distinct concentric zones of growth in its fibrous fruiting structure; perithecia appear in a single layer just below the outer crust. The fruit bodies, or stroma, appear as hard hemispherical cushions up to 4 cm in diameter on dead trunks and logs of wood, sometimes on *Antiaris africana*. The feeding mycelium in the dead wood is perennial, but the stroma remains alive for only a single season. At first it appears reddish-brown, but soon changes to black. The surface is somewhat shiny, smooth and is dotted with minute pores formed by the ostioles of the perithecia. In a vertical section, the stroma shows distinct concentric zoning of its hyphal tissue caused by the regions of thick-walled hyphae alternating with less thick regions. According to Ingold (1960) these concentric rings are in fact daily zones, and their number probably indicates the time involved in the formation of the stroma.

Just below the hard outer crust, about 250 μm from the edge of the stroma, small perithecia are crowded in a single layer. These are rather elongated, possess a conical neck and are immersed in stromatal tissue. The ascogonium is situated in the basal cone of the perithecium, but having lost its dense protoplasmic contents it is very difficult to observe. Long ascogenous hyphae arise from the ascogonium and climb up just within the perithecial wall to terminate into asci at their apical ends. The asci within the perithecium are immersed in mucilage which serves as a local water reservoir for the asci. Asci cylindrical, 70–150 × 8–12 μm, with long stalk. Ascospores uniseriate, elliptical to inequilateral, opaque at maturity, 12–17 × 6–9 μm; conidia ovate, greenish, continuous, 6–8 × 4–5 μm.

Distribution: common throughout the world on many deciduous trees.

SARCOSCYPHACEAE

BULGARIA Fries
(SARCOSOMA Le Gal)

This genus has a wide distribution throughout the temperate and tropical regions of the world. Usually the species grow in dense caespitose clusters, occasionally singly on the soil or on decaying wood of various kinds. Fruit bodies black or blackish, rather thick and very gelatinous, stipitate or sessile, externally these may be covered with hairs or may be tomentose; when dry the fruit bodies become very tough and corky, reconstituting when moist; hymenium partly or entirely covers the upper surface and forms a disc, light in colour; the entire fruit body becomes strongly wrinkled when dry; asci cylindrical or subcylindrical, eight-spored; spores ellipsoid or subellipsoid, hyaline or sub-hyaline; paraphyses slender.

Distribution: mostly in tropics and northern temperate zones.

Practical importance: saprophytic on fallen logs of hard wood; occasionally parasitic.

BULGARIA JAVANICUM (Rehm) Le Gal (Fig. 2.)
(= SARCOSOMA JAVANICUM (Rehm in P. Henn.) Le Gal)

The heavy gelatinous fruiting bodies of this species sometimes measure as much as 5 cm or more in height and over 7·0 cm in diameter. The upper portion of the apothecium is broad, tapering below to a long slender base. The outer

Fig. 2

surfaces of the fruiting bodies show distinct transverse concentric ridges, tomentose by virtue of their possession of flexuous brown hairs. Hymenium a little hollowed at first, then flat, occasionally becoming slightly convex, reddish-brown in colour, distinct, tapering to a thin upward edge of hypothecium, edge possessing flexuous brownish-black bristly hairs, 1·5 mm long. Flesh gelatinous, white, viscous, becoming watery towards the base, liquifying with age. Spores uniseriate, ellipsoid, more or less elongated, slightly asymmetric and arched, varying greatly in size, 25–53 × 13–22 μm, containing 1–3 large guttules with few smaller ones. Asci 500–550 × 17–24 μm, cylindrical, tapering below to a long slender base, eight ascopores usually lie in the upper portion leaving the lower portion hyaline to subhyaline; paraphyses numerous, filiform, 2–4 μm thick, gradually enlarging towards the apex and becoming about 6–8 μm thick, straight, septate, occasionally branched, containing brown coloured granules; marginal hairs straight, swollen at the base, 16–21 μm, thin towards the apex, septate. Flesh: internal zone filamentous with loose texture, forming a jelly-like substance; external zone pseudoparenchymatous, forming a skin-like ectal excipulum.

Distribution: mostly in the tropics.

BULGARIA GLOBOSUM Fries

This is a rather interesting fungus, its heavy gelatinous, usually large, fruiting bodies appearing singly on the ground in woods. Apothecium varies from 7–10 cm in width and 6–8 cm in height, sessile, strongly wrinkled, and brownish black; the external layer is tough, black, and consists of thick-walled hyphae covered with soft, brown hairs; the lower part of the apothecium is entirely filled with a jelly-like substance mostly yellowish-brown in colour, and causes the skin-like ectal excipulum to bulge. Shrinks greatly on drying; hymenium slightly concave as a flat layer within the apothecium sunken about 10 mm into the hypothecium and tapering to the thin inrolled margin, smooth and glistening, cracking as the outer layer dries; the base of the apothecium is connected to the substratum by means of dense tufts of mycelial strands. Asci cylindrical varying in length from 250–520 μm and 12–15 μm in diameter, tapering below; ascopores usually lie in the upper part of the ascus ellipsoid, not narrowed at ends, hyaline to subhyaline, 26 × 10 μm, occasionally with numerous guttules. Paraphyses filiform, hyaline to subhyaline, scarcely enlarged above, septate, 250–520 μm in length and 2·6–3·8 μm wide.

Distribution: tropical and temperate zones.

BULGARIA RUFA Schweinitz

This species is usually found on wood buried under the soil, growing in groups or tufts. Apothecia sessile or with a small stem, 2–3 cm in diameter, at first closed but finally opening to become shallow cup-shaped, with the margin incurved. The external surface is blackish-brown and covered with hairs; the ectal excipulum is tough and filled with a gelatinous substance; inner layer gelatinous, several millimetres thick, giving the fresh fruit body a rubbery consistency; on drying becomes coriaceous and strongly wrinkled. Hymenium closed at first finally opening, slightly concave, pale-reddish or reddish-brown; stem 1 cm long and about 4–5 mm thick, attached below by means of a dense mass of black mycelia, 7–8 μm in diameter; hairs blackish-brown, similar to those of the strand of mycelium at the base, but shorter. Asci cylindrical above, tapering below to a long stem-like base 275–300 μm in length and 12–14 μm in diameter; spores ellipsoid, uniseriate, with narrow ends, hyaline, granular within, 10–20 μm; paraphyses filiform, scarcely enlarged above.

Distribution: tropics and temperate zones.

COOKEINA Kuntze

The most conspicuous and common members of the Pezizales are included in this genus. There are three species found in the tropics, most growing on decaying wood or bark, and generally appearing during or towards the end of the rainy season. The fruit bodies grow scattered or in clusters, the open end is from 0·3 to 1·0 cm wide, stipitate, with a slender stem, to sessile, cone-shaped to cup-shaped, thin, fleshy, leathery texture. Hymenium yellow to red and of various other colours; outer surface of the fruit body paler, hispid to puberulous or glabrous; hairs, when present, composed of bundles of parallel cylindrical, septate, unequal hyphae. Ectal excipulum pseudoparenchymatous, composed of globose to polyhedral cells; medullary excipulum prosenchymatous, composed of slender, interwoven hyphae; asci suboperculate, all at same stage in development (Le Gal 1953), thick-walled, cylindrical, with a blunt and rounded base and a slender stem; spores ellipsoid, finely wrinkled or smooth, apiculate, containing oily drops; paraphyses slender, with septa, branched, occasionally anastomosing.

Distribution: tropics.

Practical importance: some species are reported to be edible.

COOKEINA SULCIPES (Berk.) Kuntze

The pinkish cups of *Cookeina* are frequently encountered growing on the dead twigs and branches of trees during the rainy season.

Apothecia solitary to clustered, 1–6 cm wide and 1–8 cm deep, cup-shaped, stipitate, rarely sessile, colour varies from chocolate brown to white, mostly of varying shades of pink or orange; external surface covered with fine, inconspicuous hairs which often form several (3–4) rings at the upper end of the cup; hairs whitish to yellowish up to 1 mm long and 1 mm thick, fasciculate, composed of strands of thick-walled, cylindrical, septate hyphae; stipe, when present, slender, up to 5 mm thick and 6 cm long. Hymenium pink to buff, fading on drying; asci 275–325 × 10–15 μ, cylindrical, forming a blunt rounded base and a slender tail-like connection; spores cylindrical-ellipsoid, 25–33 × 14–18 μ, contain two large oil drops, wrinkled; paraphyses 1–1·5 μ thick, with slightly enlarged apices, freely anastomosing, septate, a few are thick-walled, irregular, cylindrical to fusiform, appendages which project above the surface of the hymenium.

Distribution: mostly tropics, common throughout the lowlands of Central America, also in Mexico, the Caribbean, South America, Africa and Asia.

COOKEINA TRICHOLOMA (Mont.) Kuntze (Fig. 3.)

The light yellow and often scarlet cups of *C. tricholoma* are usually encountered growing scattered, or in clusters, on rotten twigs, branches and trunks of trees. The fruit bodies (apothecia) are stipitate, up to 5 cm broad and 3 cm deep, inverted cone-shaped with the margin slightly incurved; stipe slender 1–3 cm

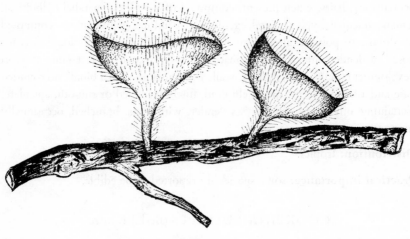

Fig. 3

long and 2–4 mm thick; hymenium pale orange to scarlet, colour fading on drying; outer surface of the apothecia paler, conspicuously hairy; hairs stiff, bristle-like, shining white to black, 2–3 mm long and 100–200 μ thick. Asci cylindrical, with blunt rounded bases 275–325 × 10–15 μ; spores ellipsoid, 28–35 × 14–17 μ and covered with fine longitudinal wrinkles, containing two large guttules; paraphyses slender, anastomosing, with the apices scarcely enlarged.

Distribution: throughout the tropics; most common on lowlands of Africa, Asia, Australia, Central America and Mexico, the Caribbean, South America, and the South Pacific area.

PEZIZACEAE

ALEURIA Fuckel

The beautiful bright coloured species of this genus have a very short life span
and appear scattered in groups or in clusters on the soil during the early spring,
after showers of rain. The apothecia are sessile or stipitate, externally smooth or
covered with delicate soft white hairs; hymenium concave or plane, even or
lacunose, usually bright orange or red; asci cylindrical or sub-cylindrical, eight-
spored; spores ellipsoid, at first smooth, becoming reticularly sculptured, often
also with apicules and ridges at the narrow ends of the spores; paraphyses slender,
usually with broad apices.

Distribution: cosmopolitan.

ALEURIA AURANTIA (Fr.) Fuckel

This beautiful cup-shaped fungus is not very common in the tropics, but may
be occasionally encountered during the early Spring growing on damp soil in
woods and, in open places, on naked clay soil. Fruit bodies gregarious or caespitose, sessile, at first globose, gradually expanding, finally reaching a diameter
up to 6–7 cm, though often smaller; shallow, cup-shaped and usually regular
in form when young, becoming irregular and often variously contorted with
age or from mutual pressure; occasionally discoid, rarely one sided, externally
whitish, pruinose; hymenium concave or almost plane, bright orange, fading
in dried specimens; asci cylindrical or subcylindrical attaining 175–250 μm in
length and 12–15 μm in thickness; spores uniseriate, arranged obliquely in the
ascus, the ends often overlapping, at first smooth with two large oil drops,
finally becoming sculptured with shallow regular reticulations, apiculate,
9–10 μm × 18–22 μm; paraphyses strongly and rather abruptly bulging at
their apices, the ends often sub-globose, filled with orange granules.

Distribution: tropics and temperate zones.

DACRYMYCETACEAE

DACRYOPINAX Martin

The species of this genus are commonly found growing on the decaying stumps of wood, in shady and moist environments. The mature specimens can easily be identified by their stipitate, spatulate and petaloid form and the characteristic unilateral inferior hymenium. Fruit bodies discoid at first, becoming stipitate; pileus obliquely cupulate, spatulate, petaloid, occasionally lobed or somewhat morchelloid; stipe covered with a layer of cylindrical, thick-walled hyphae; hymenium unilateral; inferior, smooth, forming gill-like folds or occasionally lobed; cortex stiffly gelatinous to cartilaginous; homogeneous, hyphae fibrous, thick-walled, with bulbous septa; basidia cylindrical at first, then clavate, finally becoming bifurcate; epibasidia one-spored; spores curved, cylindrical, uni- to tri-septate, germinating by conidia.

Distribution: mostly tropics, widely distributed in Central and South America, Africa.

Practical importance: saprobes on wood.

DACRYOPINAX SPATHULARIA (Schw.) Martin

This species is widely distributed throughout the tropical zones of the world. It grows on dead and decaying wood and is one of the commonest gelatinous fungi of Nigeria. The fruit bodies, erumpent, usually gelatinous when wet, drying to a horny consistency but soon reviving when moistened. Pigmentation is variable, yellow and orange predominating, but on drying the colour changes to apricot yellow. There is also a considerable variation in external appearance, some fruit bodies are simply spatulate with a tomentose stalk, while others are flabellate with a lobed and contorted hymenium. Microscopically, these variants are quite similar in appearance; hyphae fibrous, thick-walled; basidia cylindrical to club-shaped, finally becoming bifurcated; epibasidia one-spored; spore print light ochraceous buff, spores very pale yellow under the hand lens, elliptical to kidney-shaped, 4–5·2 μm × 7·4–10 μm.

Distribution: tropics, America, Pacific islands, Australia, Africa, India and West Pakistan.

TREMELLACEAE

TREMELLA Dill. ex Fries

This is a very common genus whose species are usually found growing in the woods during the rainy season. Fruit bodies are usually gelatinous, drying to a horny consistency. Many species are with a lobed and contorted hymenium. Pileus orange, brown, or white in colour, gelatinous, or varying from waxy-gelatinous and sub-fleshy (inside) to soft gelatinous; probasidia globose, oval or pyriform, becoming longitudinally or occasionally irregularly septate, four, rarely two- or three- celled, each cell producing a tubular epibasidium, sometimes notably inflated at tip below the sterigmata; basidiospores hyaline to yellowish or brownish; white, yellow, orange or yellow brown when seen in mass, globose to broadly ovate, sometimes depressed ventrally, rarely sub-allantoid.

Distribution: cosmopolitan.

Practical importance: most species are parasites on trees, some grow saprophytically on dead branches; few are edible.

TREMELLA FUCIFORMIS Berkeley

This beautiful dull-white jelly fungus grows on dead and decaying logs or stumps of trees. Fruit bodies foliaceous, the lobes caespitose, repeatedly forked or lobed, or with margin incised or crenate or sometimes entire, crisped, undulate, gelatinous, white, drying to a pale yellow; in section the lobes are usually thin, the hymenium forming a compact, amphigenous surface layer, hyphae 1–3.5 μm in diameter with clamps, sometimes nodose or with the wall irregularly thickened or with thick gelatinous sheaths; probasidia borne on a network of short, broad cells, obovate to oval or subglobose, becoming longitudinally cruciate septate, 10–18 × 7–12 μm; epibasidia, 1.5–3 μm in diameter, mostly under 30 μm in length; basidiospores ovate, flattened adaxially, 7–9 × 4–6.5 μm, germinating by repetition; conidiophores, when present, borne on the same hyphae as the basidia, the conidia forming capitate clusters, oval to subglobose, about 2–4 × 2–3 μm.

Distribution: mostly tropics: reported from Brazil, Chile, Philippines, Singapore, Africa, Australia and North America.

AURICULARIACEAE

AURICULARIA Bull. ex Merat

Fruit bodies, when fresh, rubbery-gelatinous, varying from resupinate with free margins to substipitate or occasionally stipitate; saprobic; solitary or gregarious to imbricate-caespitose; mature specimens 1–2 mm in thickness, mostly from 5–6 mm to 8–10 cm or more in width. When dry, thin, translucent to opaque, horny and brittle, upper surface pilose, hairs 65–500 μm long; lower surface bearing the hymenium, which is externally glabrous to pruinose, often with venulose folds; colour pallid or rosy to dark brown or black. Longitudinal sections show characteristic hyphal organisation with discernible zonation; hymenial layer dense; basidia cylindrical to clavate, transversely septate, giving rise to slender epibasidia terminating into stigmata. Paraphyses branched, slender; spores white to ochraceous when seen in mass, cylindrical to allantoid, germinating by a germ tube.

Distribution: tropical, subtropical and temperate regions of the world.

Practical importance: some of the species are edible: *A. polytricha* (Mont.) Sacc. is cultivated in China and cultured on poles of *Quercus*. Others parasitic.

AURICULARIA POLYTRICHA (Mont.) Saccardo

Widely distributed throughout the tropical and subtropical regions of the world and the fruit body can be easily identified in the field by its densely pilose upper surface, strongly cupulate form and dark smooth hymenium. It is a species of a very peculiar consistency, rubbery-gelatinous when fresh and drying to a brittle, cartilaginous sheet-like substance; pileus about 6 cm wide and 1 mm thick, drying brownish to almost black; longitudinal section of the pileus shows it to be divided into eight hyphal zones: or *Zona Pilosa*, hairs about 45 μm long, hyaline, forming dense tufts, with a prominent central strand. *Zona compacta* about 2 μm wide, densely compacted, individual hyphae not distinguishable. *Zona sub-compacta superior* about 85 μm wide, hyphae mostly perpendicular to the surface and about 2 μm in diameter. *Zona laxa superior* about 26 μm wide. *Medulla* about 250 μm wide, hyphae parallel with the surface. *Zona laxa inferior* about 250 μm wide. *Zona sub-compacta inferior* about

100 μm wide. Hymenium 80–90 μm wide; basidia cylindrical, becoming tri-septate, 50–60 × 4–5 μm; spores curved, cylindrical, 12–15 × 5–6 μm.

Distribution: tropical America, Africa, Australia and Pacific islands.

Practical importance: edible; a highly prized item of Chinese cooking.

AURICULARIA AURICULA (Hook.) Underwood

The species commonly called the 'Jew's ear', because of its fancied resemblance to a human ear. It is uncommon in the tropics but undoubtedly occurs abundantly throughout the temperate regions of the world. The jelly-like fruit bodies are tough when fresh, commonly solitary, occasionally gregarious or caespitose, becoming variously convoluted upon maturity; yellow brown to reddish brown when moist; sessile to substipitate; up to 12 cm wide and 0·8–1·2 mm thick. Longitudinal section is divided into six zones (Lowy 1952): *Zona pilosa*, hairs 85–100 μm long and 5–6 μm in diameter, hyaline, without central strand, rounded at tip, not in dense tufts; *Zona compacta*, 65–75 μm wide, hyphae densely compacted, individual elements not distinguishable; *Zona sub-compacta superior*, 115–130 μm wide, hyphae about 2 μm in diameter, forming a dense network giving the zone a somewhat coarsely granular appearance; *Zona intermedia*, 285–300 μm wide, hyphae 1·5–2 μm in diameter, mostly horizontal in orientation with numerous small interstices; *Zona sub-compacta inferior*, 100–120 μm wide, hyphae about 2·5 μm in diameter, forming a densely compact layer; hymenium about 150 μm thick; basidia 50–60 × 5–6 μm, cylindrical; spores allantoid, 13–15 × 5–6 μm.

Distribution: mostly in temperate regions of North America and Europe; infrequently occurring in the tropics.

Practical importance: sometimes parasitic.

THELEPHORACEAE

CYMATODERMA Junghuhn

Fruit bodies of this genus are generally found growing on wood, on dead stumps, trunks, fallen branches, roots and on the ground under trees, apparently arising from buried wood. Cap leathery in texture, semi-circular, fan-shaped, infundibuliform or pseudoinfundibuliform; closely growing fruit bodies becoming confluent. Upper surface of the cap ornamented with radiating sharply edged ridges which, in some species, are obscured by a thick tomentum. Hymenial surface irregularly folded; stipe lateral or eccentric, well developed, or reduced to a basal tubercle, tomentose; hyphae are of two or three kinds: generative hyphae, either thin-walled, hyaline, branched or thick-walled with lumina often obliterated, subhyaline, less freely branched; and skeletal hyphae, very thick-walled with lumina obliterated, hyaline, subhyaline or pale brown, unbranched, lacking clamp-connections, and narrow; binding hyphae thick-walled, with numerous short branches, lacking clamp-connections; cystidia occasionally present as apically encrusted metuloids of varying shapes from clavate, lanceolate to subglobose. Gloeocystidia thin-walled, elongated, deeply staining, frequently with a swollen base and a narrow neck; basidia clavate, usually four-spored; spores thin-walled, hyaline, non-amyloid, varying in shape from broadly elliptical to subglobose, narrower elliptical to subcylindrical, or broadly elliptical to elliptical.

Distribution: throughout the tropics and subtropics.

CYMATODERMA ELEGANS Junghuhn

One of the handsomest sporophores of the tropics, usually found growing on dead stumps, trunks and fallen branches. Fruit body up to 13 cm in length and 14 cm in diameter, and leathery in texture, fan-shaped, rarely like a funnel, usually appearing to lack one half with a short, lateral, or long, pseudocentral stipe. The entire fruit body is completely covered with a thick layer of soft matted hairs. The surface of the cap is furnished with radiating knife-edged ridges, which are usually hidden from view by the felt-like tomentum, their existence indicated only by the strikingly wrinkled appearance of the cap. The colour of the fruit body varies from pure white to fawn in fresh material, to snuff brown in very old specimens. The hymenial surface is ornamented with rather prominent, branched, radiating folds, which are united behind into 'main channels' radiating from the stem. In fan-shaped fruit bodies, the stem is short and rudimentary, but in funnel-shaped fruit bodies it tends to be well developed, often up to 5 cm long and 1 cm thick. Hyphae are of two kinds: generative hyphae, which are thick or thin-walled, septate, twisted, ribbon-like with clamp-connections at the septa; and skeletal hyphae which are very thick walled, unbranched, lacking clamp-connections and have elongated, narrow, obtuse apices. There is a distinct cuticular layer. The surface tomentum is formed of elongated, thick-walled and unbranched hyphae bearing clamp-connections. Cystidia clavate; gloeocystidia often abundant; basidia clavate, usually four-spored; spores smooth, hyaline, broadly elliptical, 6–9 × 3·5–5 μm.

Distribution: tropical Africa and Asia, and certain Pacific islands.

STEREUM Pers. ex Gray emend. Boidin

These fungi are commonly found growing on decaying stumps and branches, and on various other woody hosts. The fruit bodies are fan-like or petaloid, with a short rudimentary stipe, often exuding a red latex when bruised; coriaceous, generally dimitic but monomitic and trimitic forms are also known; context with horizontal, narrow, interweaving hyphae. Generative hyphae usually thin-walled, often slightly thicker near the growing region; skeletal hyphae start at the surface of the context and run periclinally, then raised up obliquely to finally become anticlinal and form pseudocystidia. On the upper surface of the context the hyphae are agglutinate, forming a usually yellowish crust inspected under the microscope. The surface tomentum is formed from abundant thick-walled hyphae; basidia narrow, subcylindrical, four-spored, occasionally sterile with short flabby digits, spore print white; spores oblong, subcylindrical, occasionally depressed, thin-walled, smooth, binucleate and amyloid.

Distribution: cosmopolitan.

Practical importance: most species are lignicolous and saprophytic, some growing as parasites on economically important plants.

STEREUM LOBATUM (Kunze ex Fr.) Fries

Mostly grows saprophytically on dead and decaying wood. Fruit bodies flabelliform, large (about 3–10 cm) and have long rays; at the point of attachment to the substratum the carpophore is much reduced and generally becomes substipitate; stipe flat, dark-brown, the under surface about 1 cm, sterile and therefore quite distinct from the hymenial surface, where the upper surface of the stipe is not easily differentiated from the rest of the pileus. Pileus tomentose, ochraceous, occasionally cinnamon or chestnut brown, faintly furrowed, divided into a number of concentric zones, showing shining silky bands of bay, brownish red, and greyish brown colours. Hymenium sub-smooth or mat, pruinose, bay coloured, becoming orange at the margin, and finally when old becomes hazelnut or chestnut brown. Margin thin (150–300 μm), entire or lobed and therefore slightly confluent. Flesh pale, thin; context dimitic, pseudocystidia large, 6–9 μm; pseudo-acconthophyses 18–25 μm × 3–4 μm; upper crust with much-branched hyphae; tomentum hyphae large, 4–8 μm, thick-walled. Basidia four-spored; spores smooth, oblong, subcylindrical, slightly depressed, 4·8–6·2 × 3·0–3·5 μm, thin-walled, amyloid.

Distribution: tropics; mostly in Congo, Uganda, Philippines and Nigeria.

Practical importance: cause of the decay of sapwood.

PODOSCYPHACEAE

PODOSCYPHA Patouillard

The sporophores of these fungi are either found growing in or on the wood they emerge from the soil growing from buried wood. These are usually thin and leathery in texture and of various shapes, most commonly these either look fan-like or have the appearance of a funnel. The fruit bodies, growing closely, have a tendency to grow into one another and often produce rosette-like fructifications. The margin of the pileus may be entire or delicately toothed, and sometimes appears as if it is cut into bands. The upper surface of the sporophore is either smooth or downy and almost every shade of brown, but in some species, the colour ranges from ochraceous-tawny to white, and in others the surface of the pileus is divided into a number of concentric zones of varying shades. The hymenial surface is usually of a pallid or cream colour. The well developed stipe is either covered with a dense felt-like mat of hyphae or minute bristles, and is frequently attached to the substratum by a conspicuous mycelial disc. Hyphal structure in most species is of two types: the generative hyphae, which are thin-walled, hyaline, branched, and with clamp-connections at the septa; and the skeletal hyphae which are thick-walled, hyaline, unbranched and without clamp-connections. The cuticle is usually absent in most species. The hymenium is thickened and possesses a number of long, thin-walled, undulating, hyaline structures known as gloeocystidia; these structures stain deeply with aniline blue in lactic acid, and usually have a swollen base and narrow apex. The basidia are usually thin-walled but occasionally become thickened, clavate, two- or four-spored and of very small to medium size; spores smooth, thin-walled, hyaline, nonamyloid, and varying in shape from subcylindrical to ovate.

Distribution: cosmopolitan.

Practical importance: some species are reported to be parasitic on living trees.

PODOSCYPHA BOLLEANA (Mont.) Boidin

This little fungus is a common species of the tropics and grows either on the ground or on living or dead tree trunks, branches, roots or woody fruits of various trees. The sporophores are usually 2–4 cm in length, and may occasionally grow to a length of 8 cm. They grow in clusters and sometimes fuse with each other along their margins. The shape of an individual fruit body may be spatulate, fan-shaped or more or less like a funnel; sometimes the cap splits longitudinally into segments. The pileus is smooth, shines like silk and is almost translucent varying considerably in colour; the young specimens are a whitish-cinnamon colour with white or pale margins and usually exhibit alternating light and dark concentric zones. The older specimens vary from yellowish-brown to very dark chestnut brown according to their age. The hymenial surface is smooth, and varies in colour. The pilei narrow downwards into a short or long stipe, often rudimentary, several stipes often arising from a common rooting base. The surface of the stipe is covered with a brown or pale fawn felt-like mat of hairs. When growing on wood the stem is attached to the substrate by a small basal mycelial disc. Hyphal structure consists of both generative and skeletal hyphae. Terminal or intercalary chlamydospores, produced by generative hyphae, are found scattered throughout the context and sub-hymenial tissue of the sporophore. These vary considerably in size from 6–16 × 5–10 μm and are from spherical or egg-shaped to pear-shaped. The cuticular layer is not very distinct. Pileocystidia are present, and the hymenium is thickened; cystidia absent; gloeocystidia long, undulating with swollen base and obtuse apex; basidia four-spored; spores thin-walled, hyaline and broadly elliptical to ovate in shape, the size ranges from 4–6 × 2–4 μm.

Distribution: tropics; most common in West Africa.

PODOSCYPHA PETALODES sub-species *rosulata* Reid.

The rosette-like fructification of the genus *Podoscypha* is commonly seen in tropical and sub-tropical regions. It usually grows on the ground from roots and buried wood, occasionally occurring on rotting branches and logs of wood. The complicated rosettes are often about 10 cm tall and frequently become confluent to form even larger fructifications. Occasionally the fusion is not complete and the clustered pilei look like the tufts of a deeply divided fan or are more or less funnel-shaped structures. The colour varies considerably, ranging from ochraceous-tawny to orange-brown through purplish-brown and very dark chestnut to almost black. Most of the specimens are strongly zoned and have a whitish margin. The surface is smooth, whitish, often with a pinkish tint. The generative hyphae are very abundant and rather broad with narrower side branches. The skeletal hyphae are rather scanty and unbranched and tend to taper towards the apex. There is no distinct cuticle. The stipe bears caulocystidia similar to the pileocystidia, although usually longer, thick-walled and of a more distinct brown colour. Hymenium thickening. Cystidia absent, gloeocystidia are abundant as elongated, subcylindrical structures which taper towards the apex and become expanded at the base; basidia are clavate, four-spored, thin-walled, and occasionally become thick-walled; spores are thin-walled hyaline and broadly elliptical to ovate or sub-globose, ranging from 2·5–4 × 3·5–5 μm in size.

Distribution: widespread in tropical and subtropical regions, but especially frequent in Northern India, Pakistan, Ceylon, Malaya and Nigeria.

CLAVARIACEAE

RAMARIA S. F. Grey emend. Donk

The *Ramaria* genus belongs to the family Clavariaceae which includes some of the most beautiful of fungal fruiting bodies. These are commonly known as the coral fungi because of their erect and intricately branching basidiocarps, which resemble coral. Corner (1950) considered the type of branching and hyphal structure of the fruiting body, and the presence and type of cystidia, and colour of spores, as some of the distinguishing characters among the many genera of this family.

The genus is fairly common in the tropics, possessing the largest of all clavarioid fruit bodies, and in this respect is one of the finest genera of the Basidiomycetes. Although a number of species of this genus have been described, many new species in unexplored parts of the world, such as the Himalayas and the mountains of Africa, may remain to be discovered. The genus comprises some of the more primitive fungi which are interesting from an evolutionary point of view.

Fruit bodies are massive to small, radially branched, Polychotomous or dichotomous, generally coloured, but pallid or white in some species; flesh brittle, fibrous, coriaceous, tough or gelatinous. Hymenium thickening is present or absent, often dorsiventral and sterile on the upper sides of the branches; subhymenium composed of uninflated hyphae. The hyphae are dimitic, clamped, rather long-celled, not secondarily septate and their walls are thin or slightly thickened; some species have narrow interweaving hyphae. Cystidia and gloeocystidia are absent; basidia four-spored, not secondarily septate. Spores are pale yellow, ochraceous, cinnamon or ferruginous, ellipsoid, large to small, smooth, striate, rough, rugulose, verruculose or echinulate, generally with one to several guttulae, aguttate in a few species.

Distribution: cosmopolitan.

Practical importance: most species are edible but some are *poisonous*.

RAMARIA MOELLERIANA (Bres. et Roum.) Corner

A common fungus of the tropics it grows generally in tufts on dead wood or on the ground. The coral-shaped, yellow or ochraceous fruit bodies are up to

10 cm high, much branched, coriaceous and tough; stem 20 × 5 mm, cylindrical and slightly tapering towards the apex, slightly thickened and white tomentose at the base, arising from a pallid to white mycelium. Branches repeatedly and is dichotomous, strict, erect, axils arcuate and flattened, or acute and scarcely flattened, apices subcristate or elongate-subulate; hymenium unilateral, on the lowerside of the main branches. Hyphae long-celled, with slightly or distinctly thickened walls, particularly in the narrower hyphae. Spores 6–8 × 3–5 μm, ochraceous, smooth and ellipsoid.

Distribution: tropical Africa, tropical Asia, New Guinea and Bonin Island.

POLYPORACEAE

GANODERMA Karst. emend. Patouillard

The fruit bodies of this genus are sessile and stipitate, the upper surface of the pileus being shiny due to the presence of an amorphous waxy substance secreted by the hyphae. Hyphal system trimitic; generative hyphae with clamp connections; skeletal hyphae are of two types: arboriform, showing an unbranched basal part with a branched tapering end; and aciculiform, unbranched and usually with a sharp tip; binding hyphae slender, branched and twisting. Basidia globose to clavate, 4-spored, collapsing in dried basidiocarps; spores brown, double-walled with a dark coloured inner wall bearing an ornamentation which pierces the hyaline outer wall so that the spores appear to have a spiny surface; ovate to cylindrical, or ovate with the wall unevenly thickened, the apex being always thicker, truncated or rounded; context always with a derm.

Distribution: cosmopolitan; common in tropics.

Practical importance: parasitises economically important plants.

GANODERMA LUCIDUM (Leys ex Fr.) Karsten

Fruit bodies are usually found growing on or near the trunks of deciduous trees, especially oaks. Pileus 5–20 cm wide, more or less circular or kidney-shaped, often flat and narrow, concentrically grooved, covered with a shining blood red varnished crust, sometimes lighter and sometimes almost black, stipitate or sessile. Stipe concolorous, eccentric, up to 15 cm long, the base of which is often continuous with a fairly large irregular pseudosclerotia, deeply buried in the soil with a definite brown or brownish-red rind which is often impregnated with the fragments of stone and roots, and which is eventually connected with the roots of the host tree. Context off-white becoming dull brown, corky and fibrous, up to 1 cm thick. Hymenial surface poroid, tubes white, finally becoming cinnamon, fairly long, often up to 3 cm; pores minute, 166 μm wide; dissepiments 67 μm thick; spores ovate, warty, 11–14 $\times$ 6–8 μm, rough, brown.

Distribution: mostly in the tropics.

Practical importance: parasitises frondose trees, causing a white rot, especially in oaks.

GANODERMA PSEUDOFERRUM (Wakefield) Over. & Steiumann

The fruit bodies of this species are rather common in tropical countries and are usually found growing at the base of the trunk of trees, or a short distance from the surface of the ground; their shape and colour are extremely variable. When young they are white, fluffy and spherical; when touched, yellow or brown patches appear at the area of contact; mature fruit bodies are about 40 cm or more across; the upper surface of the pileus becomes dark brown or black, often with a violet or greenish shade, the outer parts showing a brown-red zonation. Another typical characteristic is the fluffy, white margin which is always found in old fructifications. The mycelial strands growing on the surface of the root of the host grow together when older, forming tough mycelial membranes which cover the entire surface of the infected root; these membranes, in early stages, are usually white when dry, becoming dark red when wet. The affected wood shows a wet or damp rot; no rhizomorphs are observed.

Distribution: mostly tropical.

Practical importance: parasitises rubber and other economically important trees, causing a red root rot which spreads contagiously.

FOMES (Fr.) Kickx

Fruit bodies of this genus are usually found growing on logs, stumps and trunks of living trees and may generally be recognised by the extremely minute pores, which can hardly be seen without the aid of a lens, and by their woody texture. Small at first, these perennial fruit bodies attain large size in many species, the new growth of each year adding to the outward extension of the margin and downward extension of the hymenial regions. Pileus sometimes rather tough and leathery, but usually hard and woody, sessile, bracket-like; context dimitic, firm, corky to woody, cinnamon-brown; tubes frequently stratified, usually forming a number of definite layers and can be seen clearly if the basidiocarp is cut vertically; setal structures present or absent; spores hyaline or brown, ellipsoid, up to 25 μm long.

Distribution: cosmopolitan.

Practical importance: parasitic on economically important woody plants; some species such as *F. officinalis* are reported to be of medicinal use.

FOMES LIGNOSUS (Kl.) Bresadola

A very common tropical species said to be distributed worldwide and causing root diseases of many tropical plants. Fruit bodies are frequently seen growing at the bases of infected trees or on large exposed roots or decayed stumps, especially during the rainy season. Fructifications are seldom seen in well-kept plantations. Sporophores usually annual, rarely perennial, sessile, attached by broad base, effuso-reflexed, rarely resupinate, applanate to dimidiate, usually imbricate, leathery when fresh, hard and rigid on drying, 4–24 × 3 – 10 × 0·2–1·4 cm.; colour of the upper surface is dependent upon age and the humidity of the environment; when young it is a deep red-brown with a bright yellow margin, often the top is over-grown with algae giving it a greenish colour; concentrically zonate, with thin crust, tomentose, becoming glabrous on drying; context cream coloured, corky, up to 1 cm thick. Hymenial surface bright orange when fresh, reddish-brown on drying, with up to 5 mm of sterile margin; pores round, oval or angular, from 5–10 per mm; hymenial tubes about 6 mm long, reddish-brown below, sharply defined from the creamy context and do not fade internally with age but the upper surface becomes woody textured, with close-set, narrow, concentric zones, the margin becomes the same colour and hymenial surface red-brown; basidia and hasidiospores not observed; cystidia pointed, bulbous at the base, thin-walled, 5–10 μm; hyphae hyaline, thin or slightly thick-walled, septate, branched, 3·4–4·6 μm. Rhizomorphs firmly attached to infected roots, typically white and flat at the growing end, scarcely branched, often forming a white network which occasionally forms a whitish covering; as they grow older the rhizomorphs become rounded and may be tinted orange-red.

Distribution: mostly tropics.

Practical importance: parasitises rubber, oil palm, teak and other economically important plants; causes a white root-rot.

FOMES LINTEUS (Berk. & Curt.) Cooke

A common species of the tropics, its fruit bodies usually appear on living or dead trunks and branches of deciduous trees, up to 40 cm wide, applanate to somewhat ungulate, at first tomentose, the tomentum matting and then the hard upper crust becoming somewhat rimose; context silky, reddish brown, distinctly fibrous; tubes in distinct layers, soft, papery; the pores are variable in size and shape, up to 8 per mm; hyphae thin-walled, frequently septate;

setae ventricose, 13–25 × 5–9 μm; spore-print pale brown to dark reddish-brown; spores smooth, broadly oval to globose, 4–5 × 3·5–5 μm.

Distribution: mostly in tropics, Bahamas, Cuba, Jamaica, Trinidad and Nigeria.

Practical importance: Parasitic on woody trees.

FOMES NOXIUS Corner

The fruiting bodies of *F. noxius* are rather rare, but it is said that the species is distributed widely in the tropics and causes root diseases of many economically important plants. The infected roots are coated with a crust of soil particles held together by the mycelium. Pileus applanate, irregular, upper surface with a smooth, hard, brown or black crust, connate, maximum diameter about 10 cm. Context bright orange-yellow, contrasting with the dark crust; hyphal system dimitic. The larger, thick-walled, darker coloured, setae-like, unbranched and aseptate skeletal hyphae are imbedded in the usual, hyaline, thin-walled, branched and septate generative hyphae. Hymenial surface, when dry, is dark brown, black when wet, and is poroid; pores minute, darker than the context; setae numerous, rather thick-walled; spores hyaline, globose, 5 μ, not often observed. The bright colour and the unusual microscopic structure of the context are marked features of this species.

Distribution: mostly in the tropics.

Practical importance: causes a brown root-rot in Teak, Cocoa, Rubber, Sour Orange, Lime, Grapefruit, Kola, Coffee and other tropical plants.

MICROPORUS P. Benvois ex Kuntze

This genus belongs to the family Polyporaceae and includes about ten species which are closely related to each other, forming a natural group. For a discussion regarding the validity of the genus see Donk (1960). It is a fairly common genus of the tropics, the fruit bodies are usually annual, occasionally growing solitary or in small groups, all possessing a well developed central or lateral stem. Pileus funnel-shaped, conchate, flabelliform, or spatulate; stipe attached to the substratum through a mycelial disc; hymenial surface poroid; pores minute, round with even dissepiments, sometimes expanded; context of densely interwoven hyphae; hyphal system trimitic: skeletal hyphae with lumina capillary, aseptate, unbranched, hyaline, thin; binding hyphae with lumina capillary, branched, aseptate, hyaline; generative hyphae thin, hyaline, branched, septate and clamped; basidia and paraphyses forming a dense palisade which finally disorganises; basidia more or less clavate, two- or four-spored, normal; spores elliptical, smooth, hyaline, non-amyloid.

Distribution: mostly tropics; common in Africa, Central and South America, East and West Indies, China, India, West Pakistan, Philippines, Australia, New Zealand and Nigeria.

MICROPORUS XANTHOPUS (Fr.) Kuntze

This is a fairly common Xerophyte species of tropical Africa and is usually encountered growing on the bark and decorticated wood of fallen branches and on stumps of trees. Fruit bodies annual, solitary or appearing in groups, coriaceous, stipitate, funnel-shaped or often orbicular or elliptical, 3–9 cm diameter. Pileus surface concentrically zonate alternating with bands of varying shades of yellow, bay, chestnut, amber or purplish black, younger specimens are usually of lighter shades, glabrous, polished, radiately regulose; context up to 60 μm thick, chestnut coloured and of densely compacted parallel hyphae which are strongly cemented together; margin of the pileus acute, plane, brown or tipped with yellow, surface entire or crenate; hymenial surface cream, clay or ochre, decurrent, reflecting surface markings, with a sterile edge 1–2 mm wide. Stipe 1–4 cm long, 3–6 mm in width, equal or expanded apically. Hymenial pores clay coloured, 8–10 per mm, 80–120 μm in diameter and 0·5 mm deep; dissepiments 25–50 μm thick; context white, up to 1 mm thick, with interwoven compact hyphae; skeletal hyphae 5 μm, binding hyphae 4 μm and generative hyphae up to 2·5 μm in diameter; hymenial layer up to 12 μm thick with a dense palisade of basidia and paraphyses; basidia subclavate, four-spored; sterigmata erect, up to 3 μm in length; paraphyses subclavate; spores broadly elliptical, apiculate, 3·5–4 × 2–2·5 μm, smooth, hyaline.

Distribution: Africa, South America, China, East Indies, Philippines, New Guinea, Australia, India, Pakistan and Nigeria.

CORIOLOPSIS Murril

This is a common tropical genus. Fruit bodies thin, flexible or rigid, annual, epixylous, sessile, dimidiate, often largely resupinate; surface light brown, zonate, anoderm, hairy; margin thin; context thin, coriaceous to woody, pale ferruginous, sometimes almost white; hymenium concolorous; tubes small, regular, thin-walled, entire; spores smooth, hyaline.

Distribution: mostly in tropics.

CORIOLOPSIS OCCIDENTALIS (Klotzsch) Murril

This is a very variable tropical species and appears to have been described as a new species many times. It can easily be confused with other similar species, such as *C. paronius* and *Irpex maximus*, but by a careful examination of the pileal surface and the nature of its pores it can always be distinguished. (It is also very similar to some varieties of a European species *Trametes hirsuta*.) Sporophores coriaceous to corky, sessile, dimidiate, flabelliform to reniform, sometimes protracted, usually broadly, but sometimes narrowly, attached to the host, effuse-reflexed, sometimes in rosette-like clusters, rarely solitary bodies. Pileus aplanate to conchate, but sometimes with upwardly reflexed margin, 2–18 cm long, 1–10 cm wide and 0·6–2 cm thick. Upper surface of the cap tomentose, concentrically sulcate and zonate, whitish when fresh, becoming yellowish beige to ochraceous-brown to cinnamon-brown. Margin entire, thin, velvety, rigid, concolorous with the pileul surface usually bent downwards, but sometimes curved upwards. Context heterogeneous, formed by a loose, soft upper layer and fibrous, harder lower layer, sometimes sub-shining, white when fresh, then becoming cream to beige, usually with the upper layer much darker; tubes usually in one layer but sometimes indistinctly stratified at the point of attachment to the substrate, sometimes pruinate, 10–0·5 mm long. Hymenial surface poroid, sometimes subdeadaloid near the centre, white when fresh then becoming cream, dark beige or greyish; pores subcircular to elongate, large to medium, 1–3, usually 2 per mm. 250–320 (–400) × 180 × 250 μm in diameter. Dissepiments entire, obtuse and thick. Cuticle of the cap thin, 1–2 mm thick. Hyphal system trimitic; generative hyphae thin-walled, hyaline, branched with abundant clamp connections; skeletal hyphae thin-to thick-walled, usually with distinct lumen, hyaline to yellowish, unbranched non-septate; binding hyphae thick-walled to solid, hyaline much branched, non-septate. Hymenial cystidia, setae, and cystidioles absent. Hyphal pegs abundant. Basidia clavate, hyaline, four-spored; spores hyaline, cylindrical-allantoid smooth, non-amyloid, 6–7 × 2·5–3 μm.

Distribution: this is a fairly common species of the tropics and is usually found abundantly in South and Central America, West Indies, Mexico and the southern states of the U.S.A. It has also been reported from South-East Asia, Australia and Africa.

Practical importance: it is commonly found on dead wood, especially of hard wood and has been reported on *Trichilia oblanceolata*, *Bursera simaruha*, *Terminalia obovata*, and *Theobroma cacao*.

PYCNOPORUS Karsten

The first description of this genus was based on only one specimen, *P. cinabarinus* (Jacq), which occurs abundantly in temperate regions. The tropical genus *P. sanguinens* is fairly common, although the two species are often similar in some of their forms and it is not easy to distinguish one from the other. Fruit bodies annual, sometimes reviving, epixylous, sessile, dimidiate, simple or imbricate, rarely pseudo-stipitate, surface anoderm, slightly pelliculose at times, zonate or azonate, bright or dull red; context red, soft corky to punky; hymenium concolorous; tubes small, firm, thin-walled; spores smooth, hyaline.

Distribution: tropical and temperate zones.

PYCNOPORUS SANGUINEUS (L. ex. Fr.) Murril

This highly attractive species has been known from ancient times on account of its great abundance and brilliant colouring. It occurs abundantly in almost all of the tropical countries of the world and appears growing on stumps or logs of dead trees and structural timbers of almost any kind of deciduous or evergreen tree. Fruit bodies appear singly or in groups, substipitate, sessile or attenuate at the base; coriaceous when fresh and usually somewhat flexible when dry. Pileus tough, leathery to rigid, dimidiate or flabelliform, 1–7 (–10) × 1–14 (–20) × 0·2–0·5 (–2) cm; surface azonate, regulose, tomentose, to glabrous, very smooth and bright-red to flame-scarlet, fading to salmon-buff or white; margin thin and acute; context floccose, zonate, concolorous with the surface of the pileus 1–15 mm thick; hyphae of the context thick-walled, sparingly branched, yellowish under the microscope, 4–8 μm in diameter; lower surface of the pileus poroid, red; the tubes are 0·5–1·5 mm long, the mouth of the tubes being circular to angular, rather thin-walled, entire, about 2–4 per mm; dissepiments thin, 50–200 μm thick; hymenium yellowish to almost hyaline, 8–10 μm thick; basidia 6–7 μm broad, four-spored; spores hyaline to yellowish, cylindric, rarely curved, smooth, 4–7 × 2–3 μm; cystidia absent; hyphal pegs abundant; hyphae flexuous, many entirely simple, others branched, a few with crosswalls and clamps, 4–7 μm in diameter.

Distribution: throughout the tropics, common in the West Indies, North and South America and Africa.

Practical importance: it has been reported that it was formerly eaten by the locals in its young stages, but it is certainly too tough for any ordinary use in this way; causes white rot of wood and is occasionally used as a dye.

TRAMETES Fries

The genus is characterised by its coriaceous consistency and its pileus being attached to the substratum by a lateral base; trimitic hyphal system and elliptical, smooth, hyaline spores. The fruit bodies are generally annual, or in many species biennial, sometimes persisting for a few years, and leathery to corky. Pileus flabelliform, applanate, or conchate; cortex present in some species; pores round, angular, irpiciform or labyrinthiform, sometimes in 2–3 obscure strata; context white to brown, consisting of parallel or interwoven hyphae which are usually compactly-arranged; hyphal system trimitic; skeletal hyphae usually with thickened luminated walls, unbranched, aseptate, hyaline; binding hyphae with capillary lumina, freely branched, aseptate, hyaline; generative hyphae with thin, hyaline walls, branched, septate, clamped; hymenium poroid; tubes extending into an uneven depth into the cortex so that their inner terminations do not form a continuous straight line; tube mouths circular to angular, usually quite regular and entire; basidia and paraphyses forming a dense palisade, often collapsing shortly after spores are developed; basidia clavate; two-or four-spored; setae absent; cystidia present in some species; spores elliptical, smooth, hyaline, non-amyloid.

Distribution: cosmopolitan.

TRAMETES CORRUGATA (Pers.) Bresadola

This is a common tropical fungus frequently appearing on the decaying logs of wood. Sporophores sessile, applanate, dimidiate, conchate, imbricate, effuse, often reflexed to resupinate, hard, corky and rigid on drying. Pileus applanate, dimidiate, 15–18 cm long, 2–7 cm wide and 0·2–1 cm thick. Upper surface of the pileus hard, encrusted, glabrous, concentrically zonate, radially wrinkled, dark reddish-brown behind with a margin white, pallid, or cream to brownish-grey band, 0·3–0·4 cm wide. Margin thin, acute, undulate or lobed, sterile below, about 1–2 mm wide. Context homogeneous, hard and corky, showing in section thin concentric growth zones, about 2 per mm. Tubes in one layer, 1–3 mm long, concolorous with the context. Poroid surface dark cream, pores medium, circular to elliptical or subangular or somewhat daedaloid and 2–3 per mm, 195–362 μm wide; dissepiments thick and obtuse to thin and acute, entire, 70–110 μm wide, in the effused area the dissepiments become unilaterally elongated and irpicoid. Hyphal system trimitic; generative hyphae delicate, thin-walled, hyaline, branched with clamp connections; skeletal hyphae thick-walled, to solid, subhyaline to yellowish, unbranched, not septate; binding hyphae subsolid to solid, hyaline, much branched, not septate. Cystidia absent;

hyphal pegs frequent, yellowish brown, conic to cylindrical; basidia clavate, hyaline, four-spored; spores hyaline, smooth, cylindrical to ellipsoid, 9–11 × 3–4 μm.

Distribution: mostly in the tropics, especially in South and Central America, Mexico, West Indies, southern Asia, Australia, southern states of U.S.A., and Africa.

POGONOMYCES Murril

An uncommon tropical xerophytic genus which can be distinguished by its small, cylindrical, very thick-walled tubes, and its thick covering of bristly hairs. Fruit bodies usually annual, apixylous, dimidiate, sessile to flabelliform, thickly covered with rigid hairs; context dark-brown, punky; lower surface of the pileus poroid; pores small, circular; tubes short, thick-walled, light brown, mouths small, circular; spores smooth and hyaline.

Distribution: mostly tropics.

POGONOMYCES HYDNOIDES (Fr.) Murril (Fig. 4.)

A species frequently encountered in many tropical countries, it commonly grows on decaying wood. The bracket-type fruit bodies of this species are usually black, sessile, applanate, dimidiate, reniform, flabelliform, imbricate,

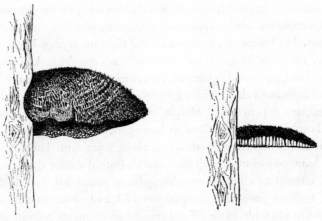

Fig. 4

coriaceous and flexible when fresh, ligneous and rigid on drying. Pileus applanate, dimidiate, flabelliform or reniform, 3–16 cm long, 2–10 cm wide and

0·5–1 mm thick, at times nodose at the base. Upper surface of the pileus hard, shiny, concentrically zonate, radially fibrous, greyish or darker, covered with a dense coat of hairs, 1–3 (–7) mm long, 200–450 µm in diameter, concolorous with the surface or entirely black, fascicled or forked, partially disorganising with age. Margin entire, undulate, crenate, usually acute or very thin, very rarely obtuse, concolorous with the pileus surface or nearly black. Context duplex, up to 1·5 cm thick, upper layer hard, radially fibrous, at times marked with concentric zones, shiny in radial section, lower layer (lower stratum) dull, up to 0·4 mm thick, medium-brown. Tubes in the several distinct layers 1–5 mm long, concolorous with the lower layer of the context. Hymenial surface poroid, olive-brown or lighter, pores medium, 154–193 µm in diameter, circular, 3–4 pores per mm. Dissepiments always thick, entire, obtuse. Upper surface of the pileus formed by a complex covering, or a dense coat, of hairs composed of tufts of yellowish-brown parallel skeletal hyphae, starting at the surface of the context and running periclinally, then raised up obliquely then becoming anticlinal; surrounding the base of the hairs there exists in the context a differentiated plectenchyma formed by shortly branched, loosely interwoven hyphae of two kinds: the generative hyphae, thin-walled, hyaline, with clamp-connections; and the binding hyphae, thick-walled to subsolid, yellowish and without clamp connections. Context formed of three kinds of hyphae: the generative hyphae, thin-walled, hyaline, branched with clamp connections; the skeletal hyphae, thick to very thick-walled but with an always distinct lumen, not branched, no clamps; the binding hyphae, thick-walled to subsolid, much branched, not septate. Hymenium without setae or cystidia, hyphal pegs usually conical more rarely cylindrical; basidia clavate, hyaline, four-spored; spores smooth, hyaline, ellipsoid when young to cylindrical when mature, non-amyloid 8–10 × 3–4 µm.

Distribution: mostly tropics, common in South and Central America, West Indies, northern states of the U.S.A., rarely in Africa.

Practical importance: it has been reported to produce white decay in wood, and frequently appears growing on *Samanea saman*, *Myristica surinamensis* and *Theobroma cacao*.

DAEDALEA Pers. ex Fries

This is a small genus, composed of about half a dozen species, and can easily be recognised by the daedaloid hymenium. It appears usually growing on wood. Fruit bodies annual, sometimes reviving for several seasons in some species. Pileus sessile or effused-reflexed or, in some species, sometimes, narrowed into a stem-like point of attachment, coriaceous to corky or very firm corky, white or brown in colour. Hymenial tubes never layered, homogeneous with the context of the pileus and not forming a distinct stratum; the mouths of the tubes elongated or daedaloid in outline, often poroid or nearly lamellate, usually entire, sometimes toothed; context of the pileus composed of hyphae with hyaline walls; hyphal system trimitic; skeletal hyphae aseptate, unbranched, with lumina almost capillary; binding hyphae freely branched, branches tapering, aseptate, lumina capillary; generative hyphae branched, septate with clamp connections, with thin and hyaline walls; hymenial layer a dense pallisade of basidia and paraphyses; in some species the ends of the skeletal hyphae, paraphysate hyphae, or cystidia project into the tubes; basidia subclavate, two- or four-spored; sterigmata present; cystidia never conspicuous, usually absent; setae absent; spores oblong or oblong-ellipsoid to cylindrical, never globose, smooth, hyaline, non-amyloid.

Distribution: cosmopolitan.

DAEDALEA ELEGANS Spreng ex Fries

This is one of the commonest bracket fungi of the tropics, often found growing on rotten logs and trunks of fallen trees in moist places. Fruiting structure sessile or with a short lateral stem, applanate, dimidiate, flabelliform, corky and flexible when fresh, firmer and more rigid when dry, up to 27 cm long (sometimes larger specimens have been observed); upper surface finely velvety to glabrous, occasionally concentric zones at the margin, smooth or with scattered small nodules; on drying it may become grey or yellowish and sometimes a purplish black; margin of the cap entire, thin, acute, sterile below; context homogeneous, hard and corky, in section showing some growth zones; hyphal system trimitic; generative hyphae thin-walled, hyaline, branched, with clamp connections and stain well in lactophenol in cotton blue; skeletal hyphae thick-walled subhyaline to pale yellow, unbranched, without clamps; binding hyphae thick-walled to solid, subhyaline to pale yellow, branched, without clamps. Hymenial surface white when fresh but becoming darker on drying; hyphal pegs setae and cystidia absent; paraphyses thin-walled, hyaline, branched, projecting over the basidial level; basidia clavate, hyaline, four-spored; basidiospores hyaline, smooth, thin-walled, elliptical to cylindrical, non-amyloid, 5–6·5 × 2–3 μm.

Distribution: all over the tropics, common in South America, Central America, West Indies, Mexico, southern states of the U.S.A., southern Asia and Africa.

Practical importance: causes the decay of sapwood; grows on a number of hosts, commonly on the leguminous and moraceous species.

POLYPORUS Micheli ex Fries

Polyporus can be identified easily by the presence of little hollow tubes on the lower surface of its cap. They generally grow on wood, rarely on the roots of herbaceous plants. The fruit body is either pleurotoid or with a central stipe, true stipe often absent and then the cap is attached directly to the substratum. Stipe generally solid, sessile; the fruit bodies possess a clearly circumscribed disk-like base; upper surface of the cap often fibrilose, squamose, pilose, or radially lineate-striate. The cap may be hygrophanus or not, its cuticle not separable. Tubes with thin walls, often radially elongated in perpendicular section, varying from small to very wide, and then often alveolar with angular or round pores. Context fleshy, toughish to very tough, and in dried specimens almost woody, colour white or pale. Hyphae hyaline, many of them conspicuously thick-walled, inamyloid, with clamp connections; cystidia mostly absent or inconspicuous, often branched; hymenial pegs often conspicuous; basidia clavate, four-spored; spores mostly cylindrical, ellipsoid-oblong, rarely fusiform, hyaline, smooth, thin-walled, inamyloid.

Distribution: cosmopolitan with the exception of arctic and antarctic zones.

Practical importance: many species are parasitic on fruit trees; some are edible and are sometimes sold in local markets.

POLYPORUS DERMOPORUS Person
(= FAVOLUS BRASILIENSIS Fries)

A common species of tropical America and Africa, it grows on dead branches of deciduous trees, and is often observed growing on *Celtis*, *Quercus* and *Vitis*. Fruit bodies are white, fleshy and usually attached to the substratum by a narrow stem-like base, or else stipitate, tough when fresh, rigid and brittle when dried. Pileus reniform to flabelliform, often clustered, 4–8 cm wide, 1–5 mm thick, short hispid-pubescent at the base or entirely glabrous, azonate, more or less striate and occasionally tesselate; margin thin, usually deflexed on drying, entire to fimbriate-ciliate, sterile, upper surface of the pileus smooth, transluscent; hymenial surface poroid, white to cream-coloured, mouth elongated and hexagonal, coalescing radially at times; tubes thin-walled, radially elongated, 1–3 mm long and 1–3·5 × 0·5–2 mm wide, thin-walled, sometimes denticulate or fimbriate; context white, rather soft when fresh, tough and almost woody when dried, reviving again on remoistening, up to 2·7 mm thick; stipe distinct and central, to indistinct and lateral, 1–3 cm long, 2–10 mm thick, marked with decurrent tubes, pubescent to tomentose; spore print pure white; spores short, cylindrical, smooth, hyaline 8–11 × 3–4 μm, thin-walled and inamyloid; basidia broad, clavate, four-spored; cystidia absent; trama of the pileus inamyloid; hyphae with clamp connections, forming a homogeneous tissue except for many vesicle-like openings 15–26 μm in diameter, extremely sinuous and irregular, thin-walled, sparingly branched.

Distribution: mostly in tropical America and Africa.

POLYPORUS ARCULARIUS (Batsch) ex. Fries

This species usually grows solitarily in small groups on decaying logs lying on the ground, and appears during the rainy season. Fruit bodies annual, coriaceous, somewhat brittle when dried and are recognised by the central stem, small horny cap with ciliate margin and large angular pores. Pileus orbicular, umbilicate or infundibuliform, 1–8 cm in diameter, 1–4 mm thick; surface azonate and depressed in the centre, squamulose, hispid, tomentose or glabrous, cinnamon buff to antimony-yellow when fresh, on drying it becomes brown; margin acute, ciliate and incurved when dry; cuticle thin, consisting of parallel compacted hyphae; hymenial surface decurrent, concave, white to pinkish-buff, on drying light pinkish-buff; tubes 1–3 mm long, mouths large angular, or polygonal, about 1 mm broad; dissepiments hyaline under the microscope, 150–250 μm broad, tapered towards the mouths to an acute edge, toothed when old; stipe central, slender, 2–7 cm long and 1–3 mm thick, solid, squamulose, hispid tomentose or glabrous above, fibrillose and bulbous at the base, concolorous with the pileus or slightly darker; context 0·1–0·5 mm thick with densely woven hyphae; skeletal hyphae up to 9 μm in diameter, freely branched, without cross walls; generative hyphae thin-walled, branched, septate and with clamp connections; hymenium hyaline; basidia sub-clavate or cylindrical, densely packed in permanent palisade, four-spored; spores hyaline, smooth, elongated, ellipsoid, apiculate, 7–9 × 2–3 μm; hyphal pegs present, usually incrusted.

Distribution: cosmopolitan, mostly in Europe, North and South America, Australia, New Zealand and Africa.

Practical importance: associated with a white rot.

PLEUROTUS (Fr.) Quélet

The white or pigmented fruit bodies of these fungi are a familiar sight in most parts of the world; the species are usually encountered growing on wood, more rarely on the other plant tissues, and on dead or living hosts. The genus is rather well known. However, the intraspecific taxonomy of certain groups is still in need of further organisation. Hymenophore lamellate; hymenophoral trama completely irregular. Spore print pure white or cream. Spores hyaline, smooth, thin, cylindrical, non-amyloid; basidia normal; cheilocystidia usually present. Subhymenium strongly developed, well differentiated and broad. Stipe present, more rarely absent. Context fleshy to tough. Hyphae with numerous clamp connections.

Distribution: cosmopolitan.

Practical importance: the genus contains some of the most valuable edible mushrooms. Occasionally these parasitise on living trees. The sclerotium of *P. tubber regium* serves the natives for food and medical purposes. (See Fig. 6.)

PLEUROTUS SQUARROSULUS (Mont.) Singer (Fig. 5.)

Widely distributed in southern parts of Asia and is clearly a common species in tropical west Africa. Fruit bodies develop in large numbers as groups or tufts on fallen trees, logs of wood and on wooden poles. The cap ranges from 2·0 to 13·0 cm in diameter, circular or shell-shaped when mature, depressed in the centre, often infundibuliform, fleshy when fresh, hard and rigid when dried. Upper surface squarrose; margin thin, regular, bent-down and sometimes rolled inward; olive-brown scales can be observed in young specimens, and are composed of thin-walled, septate hyphae with clamp connections. Gills decurrent, crowded and white. Stipe 0·5–6·0 cm long, 0·2–1 cm wide, central, lateral or eccentric, occasionally branched, equal, hollow, squarrose, annulus absent. Context two layered: outer layer has white cap and stem; inner layer yellowish-white, fleshy-coriaceous, with inamyloid, non-gelatinised, with a dimitic hyphal system of thin-walled generative hyphae with clamp connections, and thick-walled binding hyphae with numerous side branches of limited growth. Hymenophoral trama irregular, hyphae branched, hyaline, thick-walled. Subhymenium narrow. Cystidia clavate, thick-walled; basidia clavate, two- to four-spored; hyphal pegs numerous, consisting of parallel thin-walled hyphae, spores 4–9 × 2–4 μm, cylindrical, hyaline, inamyloid, thin-walled; spore print cream.

Distribution: tropical and subtropical zones. Most common in India, Ceylon, Burma, Ivory Coast, Ghana and Nigeria.

Practical importance: edible, but tough. Used in soups by local people.

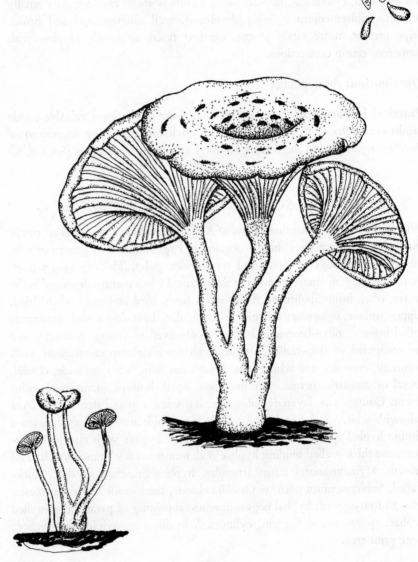

Fig. 5

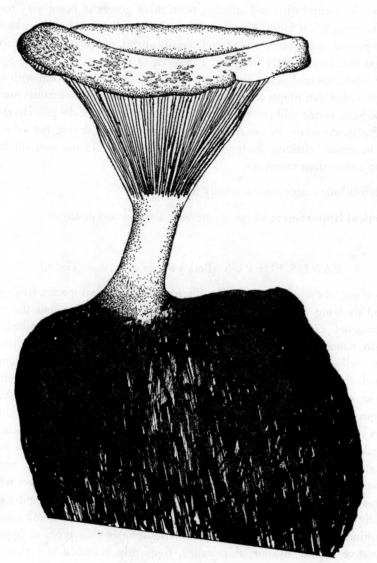

Fig. 6

PANUS Fries

The fruit bodies of this genus usually grow on wood, pleurotoid in habit, but often with central stipe and differing from other genera in being very tough and reviving when moist. Pigment present, but usually not brightly coloured. Stipe present, more rarely absent; veil usually absent, gills decurrent; hymenophoral trama completely irregular, consisting of thick-walled hyphae. Spore print white; spores hyaline, smooth, nonamyloid, always cylindrical with very thin to rather thin simple walls; basidia normal; hyphal pegs sometimes present in the hymenium; subhymenium hardly noticeable; edge of the gills lacerate – denticulate-crenulate, or entire. Context very tough on drying but reviving and becoming chamois leather-like on remoistening. Trama non-amyloid; clamp connections numerous.

Distribution: cosmopolitan, mostly in tropics.

Practical importance: all species are very active wood destroyers.

PANUS FULVUS (Berk.) Pegler & Rayner (Fig. 7.)

This is one of the most common and beautiful pan-tropical species, frequently found growing on the decayed logs of wood almost throughout the year. It possesses extremely variable characters and has been described by various names. It can easily be identified by its tough texture, velvety stipe and often well developed pseudosclerotium. Cap 2–10·5 cm in diameter, funnel-shaped, thin, tough, very light yellowish-brown to dark brown, occasionally zonate, ornamented with very fine smooth short, erect, concolorous hairs; margin thin, involute in dry conditions, reviving when moist, ciliate, sometimes cracked. Gills decurrent, lighter than the cap, becoming darker, densely crowded, thin, narrow, coriaceous. Stipe central often up to 11 cm long, generally smaller, slightly narrowing at the base, cylindrical, solid, tough, surface ornamented with erect hairs forming a persistent velvety layer which covers the entire stipe and terminates abruptly at the base of the gills; stipe slightly darker than the cap and usually arises from a large, smooth pseudosclerotium. Context whitish, inamyloid, consisting of two types of hyphae; generative hyphae hyaline, thin-walled, frequently branched and clamped; skeletal hyphae hyaline, occasionally pigmented, very thick-walled and un-branched. Spore print white; spores 5–8 × 2·5–4 μm, elongated, hyaline, inamyloid, thin-walled; basidia clavate, four-spored; cheilocystidia absent; gill edge sterile; pleurocystidia not abundant, cylindrical, clavate, hyaline, projecting beyond the basidia. Hymenophoral trama irregular, broad, hyaline, inamyloid, consisting of interwoven generative and skeletal hyphae. Pileal surface formed from tufts of unbranched, septate hairs, up to 1 mm long, very thick-walled, clamped and brown-coloured hyphae.

Distribution: pan-tropic.

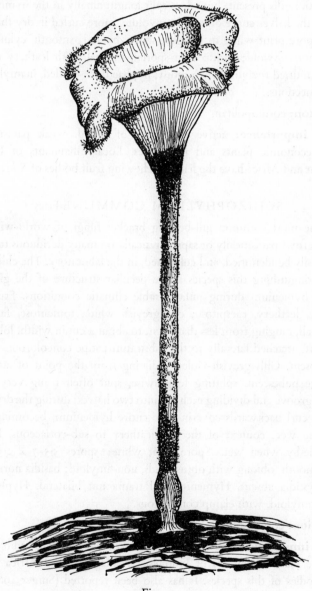

Fig. 7

SCHIZOPHYLLUM Fries

This is a very common genus identified easily by its characteristic gill structure. Fruit bodies xerophytic, generally encountered growing on wood or on other dead or living organic matter. Cap sessile, fan-shaped, hymenial surface very characteristic; gills present, but edges split longitudinally in the hymenophoral trama and the halves curve outwards, involute, more curled in dry than in wet weather, spore print white to pinkish; spores hyaline, smooth, cylindrical, or ellipsoid, non-amyloid; basidia four-spored, normal, flesh leathery to corky, tough when dried reviving when moist; hyphae thick-walled, inamyloid, with clamp connections.

Distribution: cosmopolitan.

Practical importance: active destroyers of wood, weak parasites often attacking economic plants and fruit trees. Local inhabitants of Indonesia, Madagascar and Africa have the habit of chewing fruit bodies of *Schizophyllum*.

SCHIZOPHYLLUM COMMUNE Fries

One of the most common gill-bearing bracket fungi of world-wide distribution, it grows parasitically or saprophytically on many deciduous tree species and can easily be identified, and cultivated, in the laboratory. The chief characteristic distinguishing this species is the peculiar structure of the gills which cover the hymenium during unfavourable climatic conditions. Fruit bodies xerophytic, leathery, caespitose; cap greyish white, tomentose, fan-shaped, usually small, ranging from less than 1 cm to about 4 cm in width, lobed, often deeply cleft, attached laterally to the substratum; stipe concolorous, rudimentary or absent. Gills greyish-violet, radiating from the point of attachment, thick, edge pubescent, splitting lengthwise, split often being very shallow, forming a groove and dividing each gill into two halves; during the dry weather the halves curl backwards covering the entire hymenium, becoming straight again when wet; context of the cap leathery to sub-coriaceous, becoming toughish-fleshy when wet. Spore print white; spores $5 \cdot 5$–$7 \times 2 \cdot 5$–$3 \cdot 5$ μm, hyaline, smooth, oblong with obtuse ends, non-amyloid; basidia normal, four-spored; cystidia absent. Hymenophoral trama not bilateral. Hyphae thick-walled, inamyloid, with clamp connections.

Distribution: cosmopolitan.

Practical importance: active destroyer of wooden materials; edible, used in soups. The local inhabitants of Indonesia, Madagascar and Africa chew the fruiting bodies of this species. It has also been reported (Singer 1962) that it causes a disease, basidioneuromycosis, in man.

HYGROPHORACEAE

HYGROCYBE Kummer

The species belonging to this genus can be recognised by their bright colours, and grow usually on the soil, in open fields, meadows, mountain slopes, lawns and dense forests. The cap is either viscid or dry, often bright red or bright yellow, or fire-red, more rarely violet, green, or pink, or dull coloured, or colourless. In brightly coloured forms the gills are either decurrent to adnexed, or adnexed to sinuate; in dull coloured and colourless forms the gills are usually non-decurrent. Spore print white; spores smooth, thin-walled, non-amyloid, uninucleate or more rarely binucleate; basidia two- or four-spored; cystidia sometimes present, filamentous. Hymenophoral trama subregular or regular. Stipe striate or smooth, dry or glutinous. Context mild, more rarely bitter, latex absent.

Distribution: nearly cosmopolitan.

Practical importance: nearly all species are edible.

HYGROCYBE FIRMA (Berk. & Br.) Singer

This scarlet-red species has been reported from many tropical countries. Cap convex, becoming expanded, slightly umbilicate in the centre, non-viscid, silky, with minute scales towards the central disc; margin involute and lobed; gills broadly sinuate to short decurrent, pale orange, thick; edge concolorous, entire. Stipe 1·5–4 cm long, equal or becoming wider towards the cap, hollow, pale orange-yellow, smooth with a rough base which is covered with sharp-pointed hairs, and a pruinose apex. Flesh thin, soft, concolorous with surface of stipe, consisting of loosely interwoven, thin-walled hyphae. Hymenophoral trama regular, hyaline or nearly so, consisting of thin-walled, much inflated hyphae; subhymenial layer well developed. Cheilocystidia hyaline, thin-walled, forming a sterile gill edge, varying in shape from piriform to clavate or cylindrical; pleurocystidia absent; basidia dimorphic, large or small, clavate, or clavate to subcylindrical, four-spored; spore print pale cream; spores ellipsoid, smooth, hyaline, inamyloid, dimorphic, large, 9·7–14 × 5·2–8 μm, or small 5·5–8·5 × 4–5·3 μm. All hyphae have clamp connections.

Distribution: widespread in tropics, mostly in Ghana, Nigeria, Congo, Trinidad, Mexico, Malaya and Ceylon.

Practical importance: edible.

TRICHOLOMATACEAE

OMPHALOTUS Fayod

The fruit bodies of this genus grow abundantly on decaying stumps and trunks of both coniferous and frondose trees. Cap fleshy, luminous when fresh, and brightly coloured; gills deeply decurrent. Stipe fibrous, fleshy, central or eccentric, spore print pure white; spores, when old, have slightly thickened walls, hyaline, subglobose to short-ellipsoid; basidia normal; cystidioles often numerous near the edges of the gills. Spores, basidia and hyphae non-amyloid; hyphae have clamp connections; epicutis little differentiated; hymenophoral trama irregular or subregular with a recognisably axillary trend, near the thin subhymenial layer.

Distribution: almost cosmopolitan.

Practical importance: active destroyers of timber. They are poisonous. The yellow pigment is extracted and used for cytological stains. It has been reported that the American race or species contains two antibiotics, illudin M and illudin S.

OMPHALOTUS OLEARIUS (Dc. ex Fr.) Singer

This colourful mushroom is widely distributed in the tropical and sub-tropical regions of the world. The bright apricot-orange fruit bodies usually appear in caespitose clusters on the decaying stumps and roots of both coniferous and deciduous trees. Cap fleshy, 3–6 cm in diameter; when young the margin of the pileus remains attached to the stipe via a membranous veil; when expanded, fragments of the veil are often found attached to the margin; umbonate, convex; margin incurved; cuticle brown, darker in the centre with a coating of silky fibres and brown scales, separating from the flesh when old. Stipe 5–9 cm long, 1–2 cm thick, straight or curved near the base, not bulbous, fibrous, hollow, annulus fragmentary, dark brown, disappearing with age. Gills luminescent, free, forming a ring around the stipe, moderately distant with short lamellule. Basidia clavate, four-spored; spores hyaline, subglobose to broadly ellipsoid, 5 × 7 μm.

Distribution: Africa, North America and other tropical and sub-tropical regions.

Practical importance: active destroyer of timber. Flesh toxic, severe cases of gastrointestinal poisoning have frequently been reported.

1. *Aleuria aurantia;* on the ground in woods.

2. *Cookeina sulcipes;* a group of fruit-bodies projecting from a dead log of wood.

3. *Cymatoderma elegans;* the beautiful white anastomosing sporophores growing on a log of wood.

4. *Podoscypha bolleana;* the silky and almost translucent fruit-bodies growing on a log of wood.

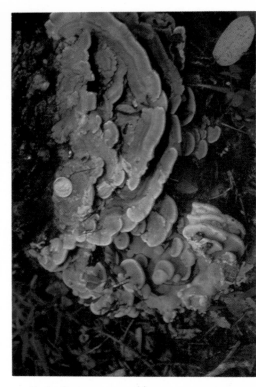

5. *Ramaria moelleriana;* on soil under shade.

6. *Fomes lignosus;* on a rubber tree stump.

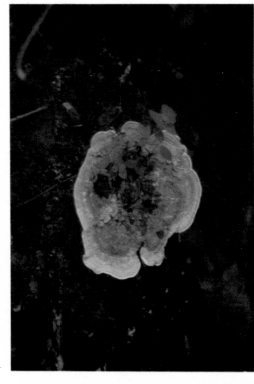

7. *Microporus xanthopus;* grows on fallen branches of deciduous trees; the colour of the upper surface varies, being lighter when young.

8. *Coriolopsis occidentalis;* an extremely common saprophyte.

9. *Pycnoporus sanguinelis;* Bright red fructifications growing on a log of wood.

10. *Deadalea elegans;* it grows on tree trunks as well as logs of wood. Note the anastomosing pores.

11. *Polyporus dermoporus;* the white fragile sporophores growing on a dead branch. Note the hexagonal mouth of the tubes.

12. *Pleurotus squarrosulus;* it grows on tree trunks, logs of wood and on garden fences; edible but tough.

13. *Tricholoma lobayensis;* edible, gathered from University of Ife, Ile-Ife Campus, Nigeria, 1969. (A mushroom hunter prepares his booty for sale.)

14. *Favolaschia thwaitesii;* a beautiful viscid orange species occurring on branches of deciduous trees.

15. *Chlorophyllum molybditis;* on rich manured soils, in forests or on lawns amongst grass, scaly cap and greenish colour of the gills. Often all parts turn reddish where touched or bruised.

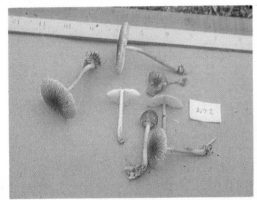

16. *Lepiota goosensiae;* upper surface coral-red, darker in the middle with fine radial striations, hymenial surface white, on soil under shade, in forests.

TRICHOLOMA (Fr.) Quélet

Species belonging to the genus *Tricholoma* are fairly uncommon in the tropics and very few species have been reported from this zone. They often favour woodlands, growing on seasonally wet ground, more rarely in open fields, and very rarely on wood. The cap is viscid or non-viscid, smooth or ornamented with delicate fibrils. If pigment is present, it is incrusted in the hyphal walls in most of the clampless species, otherwise it is dissolved in the cell sap. Cuticle consists of interwoven hyphae which are usually without clamp connections. Gills are distinctly emarginate, sinuate, thin to medium-thick. Spore print white; spores smooth, hyaline, usually very thin-walled; inamyloid, ellipsoid to subglobose; basidia normal, four-spored; cystidia usually absent. Hymenophoral trama regular to almost subregular, with parallel to somewhat interwoven, thin and elongated hyphae. Stipe central, fleshy, smooth to fibrillose, sometimes very hard but never cartilaginous or leathery, solid, or hollow; veil absent, context of pileus fleshy, mild, bitter.

Distribution: mostly in the warm temperate and sub-tropical zones, very few species truly tropical.

Practical importance: some species are edible *T. matsutake* is economically the most important species and is sold in Japanese markets in dried, as well as fresh condition, and in cans. *T. lobayensis* is a tropical species and can often be seen in local markets as dried and fresh specimens. A number of species, such as *T. saponaceum*, have antibiotic properties.

TRICHOLOMA LOBAYENSIS Heim (Fig. 8.)

This species usually appears after heavy showers during the rainy season growing on slopes and on plant debris hidden in the soil. The fruit bodies have a pleasant smell of edible mushrooms and appear in groups or clusters. Fructifications of all ages frequently arise from a single base. Stipe 15–30 cm in length, irregular, unequal, tapering towards the apex, smooth and fibrous when fresh, contorted when dried, base 4–6 cm wide, often the bases of many stipes are joined together. Pileus 10–20 cm in diameter, smooth, naked, convex, ivory white; margin thin, entire, non-striate, remaining decurved. Gills decurrent, ivory or cream colour, moderately distant with short lamellullae. Flesh white, fibrous, homogeneous. Spores $5 \cdot 5$–$6 \cdot 7 \times 3 \cdot 8$–$4 \cdot 9$ μm, smooth, hyaline, non-amyloid, appendiculate; basidia 4-spored; cystidia absent. Gill trama regular, consisting of more or less parallel thin-walled hyphae. Pileus surface non-differentiated; context homogeneous, lacunose, light, consisting of cylindrical

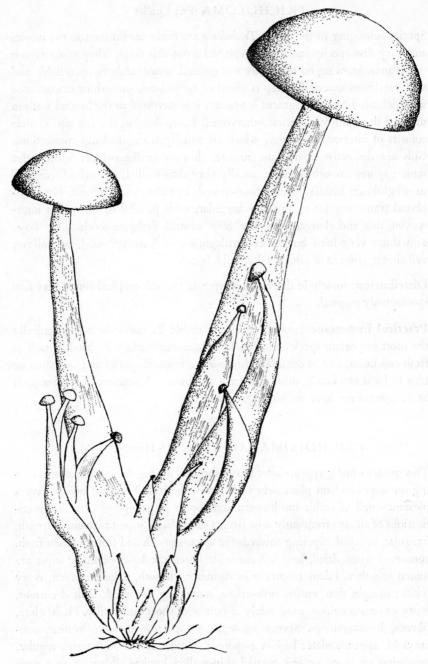

Fig. 8

interwoven 4·9–12 μm long hyphae, provided with clamp-connections. Reaction with ordinary guaiac tincture is negative, with KOH positive, with HCL negative at first then positive, the colour turning to greenish-yellow; with sulfobenzaldehyde the reaction is positive (yellow-green colour); with Melzer's reagent positive, the colour turning to reddish-brown; and with NH_3 negative.

Distribution: reported from central and west Africa.

Practical importance: edible.

ARMILLARIELLA Karsten

This genus, which includes important tree-destroying fungi, is a rather small group of fairly large, fleshy fungi encountered frequently growing on wood, among deep moss but rarely on bare soil. The cap is squamulose or smooth; gills close to distant, moderately thick, decurrent. Spore print usually pale ochraceous or pure white; spores often with thin to moderately thick-walled, hyaline, non-amyloid; basidia normal, usually four-spored; cheilocystidia frequently present; cystidia usually absent; stipe fibrous-fleshy to fleshy, annulate; context fleshy, not tough; hyphae thin- to moderately thick-walled, without clamp-connections, non-amyloid.

Distribution: cosmopolitan.

Practical importance: destructive parasites of economically important trees. It damages such crops as sweet potatoes and peanuts.

ARMILLARIELLA MELLEA (Vahl. ex Fr.) Karsten

Commonly known as 'honey' or 'boot-lace' fungus, this species usually forms clusters on and about tree stumps, occasionally solitary specimens do occur. It has a world wide distribution and is one of the commonest fungi in both temperate and tropical regions. *A. mellea* is one of the most variable species and several varieties have been described. Cap diameter 3–10 (–14) cm, fleshy, at first convex then spreading out with a thin, striate margin, typically a dark honey-yellow, but sometimes lighter, or tawny or sooty brown with a greenish tinge, flecked with brown or blackish recurved fibrillose scales which are more distinct at the centre, disappearing with age. Gills whitish, then pale brownish-yellow, often having brown spots and finally discoloured, arcuate, adnate or slightly decurrent. Stem long, fibrous, hollow, yellowish or brownish, darker at the slightly swollen base, and often becoming greenish black, with a large white membranous ring, often with a yellow border, usually persistent and later becoming discoloured; the base of the stipe is attached to the substratum through the thin, black rhizomorphs. Spore-print white, spores 7·5–9 × 5–6·5 μm, ovate, thin-walled, hyaline, non-amyloid; basidia normal, four-spored; context fleshy; hyphae without clamp-connections.

Distribution: cosmopolitan.

Practical importance: edible. A serious parasite of woody plants, including forest, ornamental and orchard trees, small fruit and vegetable bushes, shrubs, and other economically important crops such as rubber and tea. It destroys the roots and the lower parts of the trunk and spreads through the soil by means of its long, thick, black rhizomorphs.

NOTHOPANUS Singer

This genus consists of several species, the fruit bodies generally grow on dead or living wood. Habit of the fruit body pleurotoid, very rarely with a central stipe. Cap dry, glabrous or with radial fibrils; cuticle little differentiated. Gills adnate to decurrent, moderately close to distant, white. Stipe present or absent, rarely central, more often eccentric to lateral, white; veil absent. Context usually colourless, in young fruit bodies it is rather soft and fleshy but becomes tough with increasing age. Hyphal walls thin in younger specimens but become thicker when aged, inamyloid, non-gelatinised, clamp-connections present. Hymenophoral trama generally sub-regular. Cystidioles sometimes present, basidia normal, spore-print white; spores smooth, hyaline, ellipsoid or sub-globose, never cylindrical, thin-walled, non-amyloid.

Distribution: tropics and sub-tropics.

Practical importance: the *Nothopanus* species are undoubtedly wood destroyers, sometimes damaging living trees.

NOTHOPANUS HYGROPHANUS (Mont.) Singer

The fan-shaped white fruit bodies of this fungus grow solitarily or in groups on logs of dead wood. The cap sometimes grows up to 8 cm in diameter, membranous-coriaceous, uniform to flabellate, at first convex becoming flattened and depressed towards the base, typically pure white but sometimes has persistent dark purplish-brown spots on the surface near the base, hygrophanous, translucent when fresh, dry, gelatinous, radially fibrilose. Margin thin, strongly decurved, sinuate fimbriate in mature specimens. Gills decurrent, white, broad, rather thick, slightly intervened, occasionally forked towards the margin. Stipe short, usually lateral, often eccentric, rarely absent, solid, white and glabrous. Context very thin but tough and fibrous, white, inamyloid, consisting of firmly interwoven, thick-walled hyphae. Spores hyaline, thin-walled, inamyloid, ellipsoid, $4 \cdot 5 \times 2 \cdot 8$ μm; spore-print white; basidia clavate to cylindrical, four-spored; cheilocystidia and pleurocystidia absent. Hymenophoral trama, sub-regular to irregular, hyaline, inamyloid, consisting of branching hyphae. Subhymenial layer well developed, consisting of thin-walled hyphae. All hyphae with clamp-connections.

Distribution: Sierra Leone, French Guinea, Ghana, Uganda, Kenya, Tanzania and other tropical parts of the world.

MARASMIELLUS Murrill

These small and often neglected fungi grow on wood, herbaceous stems, fruits, leaves and needles, in deep moss and on other dead or live vegetation. Habit mycenoid to marasmioid rather rarely pleurotoid, and then small with a central, curved stalk. Cap thin, membranous, and more or less transparently striate, usually hygrophanous, with a specialised epicutis with diverticulate elements or hairs or dermatocystidia; gills broad to very narrow, subfree, sinuate, adnexed, adnate, or decurrent. Hymenophoral trama regular to irregular and sometimes rather mixed. Spores hyaline, smooth, inamyloid; basidia normal; gloeocystidia never present. Stipe thin, short, eccentric, curved, somewhat tough, usually with a softer stuffing, or narrowly hollow.

Distribution: almost cosmopolitan.

Practical importance: parasitic on living trees, grasses and sedges, and on economically important plants such as coffee and sugar cane.

MARASMIELEUS INODERMA (Berk.) Singer

This small, gregarious species usually grows on dead or living palm trees, and normally covers a considerable area of the trunk surface. The fruit bodies are occasionally tinted lilac. Pileus up to 2 cm in diameter, originally white becoming yellowish, sulcate-striate, convex; gills subdistant, of two or three lengths, adnate, broad, thin, not intervened; stipe short, recurved, eccentric, cylindrical, smooth, concolorous; trama thin, of loosely woven thin-walled hyphae, not gelatinised, non-amyloid; pleurocystidia absent; spores broadly elliptical, 7–10 × 4·0–6·6 μm (average of 8·5 × 5·7 μm); cystidia many, clavate, 4–30 × 6–11 μm, with numerous obtuse appendages.

Distribution: throughout the tropics, mostly in South America, Ghana and Nigeria.

Practical importance: grows on living hosts, especially palms.

MARASMIUS Fries

The species of this well-known genus frequently grow on sand, or on forest soil, fields and on lawns, occasionally growing on fallen leaves and on dead or living wood and on other plant tissues, such as stems, grass roots and bamboo twigs. The fruit bodies shrivel on drying but do not become putrescent, reviving when moist and continuing to produce spores. Pileus often membranous; epicutis consists of irregular or hymeniformly arranged elements, their apices ornamented with narrow appendages to give these elements a broomlike appearance. Spores hyaline, thin-walled, variable in size and shape, mostly cylindrical and ellipsoid, smooth, inamyloid; basidia normal; cystidia or cheilocystidia present, the cystidia on the sides of the lamellae often thick-walled. Stipe sometimes shiny, often dark-coloured, string-like or like a very thick hair, cartilaginous, sometimes attached to the rhizomorphs or rhizoid-like fibrils or small pseudorhizas, fibrous at the base, without a mycelial disc and without latex; context often consists of thick-walled hyphae with clamp-connections, pseudoamyloid or inamyloid.

Distribution: cosmopolitan, mostly found in the tropics.

Practical importance: some species parasitise economic plants of the tropics, such as tea, coffee, rubber and sugar cane. Some are edible.

MARASMIUS ZENKERI Hennings (Fig. 9.)

This beautiful species rather commonly grows in tropical forests under shade on the ground or on leaf litter, particularly in sandy soils among fallen bamboo leaves. Its colour varies greatly, often being washed out by rain. Cap 3–10 cm in diameter, convex then expanded convex, slightly umbonate, to sub-umbonate, deeply furrowed, pinkish-lilac to deep purplish-lilac particularly at the centre, frequently a dusty pink, light purplish, or lighter, over the radial ridges of the striations, smooth and rugulose over the disc, devoid of yellow pigment, smooth. Stipe (4) 12–17 × 0·2–1·2 cm, cylindrical with a swollen base, tapering towards the apex, hollow, smooth, brown becoming darker

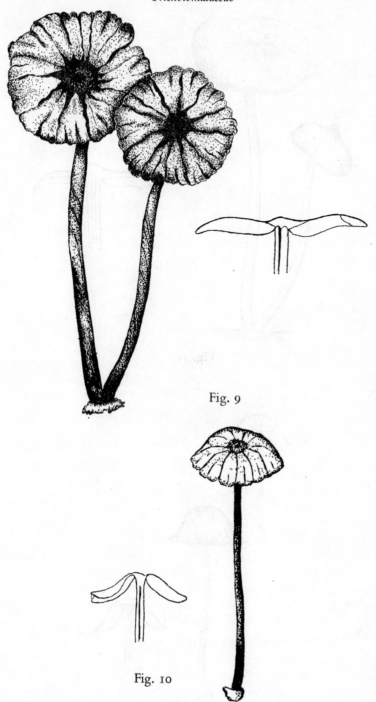

Fig. 9

Fig. 10

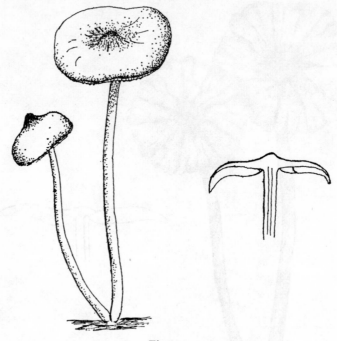

Fig. 11

Fig. 12

brown or reddish-brown at the base, often the entire stipe is soot-coloured; basal mycelium cottony, silky-tomentose, pale ochre, grey or whitish. Gills distant, slightly ventricose, narrow or large, without interveining, adnexed to free, pale lilac, edge lighter; spore-print whitish or very pale lilac; spores hyaline, 18–22·5 × 3·5–4 μm, fusiform or clavate-fusiform, thin, inamyloid; cystidia absent, cheilocystidia numerous; Pileal surface hymeniform, consisting of thin or slightly thick-walled elements; context whitish, thin and pseudoamyloid. Hymenophoral trama hyaline and regular.

Distribution: tropical Africa.

MARASMIUS UMBRINUS Pegler

This is a very distinctive species of the genus *Marasmius* and may be recognised by its very dark amber colour and the velvety texture of the cap. The beautiful small fruit body usually grows on fallen leaves during the early days of the rainy season. Pileus 4–35 mm in diameter, convex then expanded plane, thin, amber in the centre, becoming lighter towards the margin; surface smooth, velvety, covered by delicate hairs; margin radially striate. Gills free, narrow, crowded and of a snuff-brown colour. Stipe 4–8 cm in length, equal, cylindrical, hollow, umbrinous at the base becoming lighter towards the apex, pruinose over entire length, arising from a well developed, pale brown, mycelial pad. Context thin, brown, dexterinoid; spores 11·5–15 × 3·3–4·8 μ, cylindrical-clavate, hyaline; thin-walled, inamyloid; basidia clavate, four-spored; cheilocystidia present though often sparse, hyaline, or with a brown membranous pigment, with a subglobose to cylindrical base bearing thin-walled and hyaline or thick-walled and brown setules. Pleurocystidia absent; basidioles fairly abundant. Hymenophoral trama brown, sub-regular to irregular, dextrinoid, consisting of narrow, pale brown, thin-walled hyphae. Sub-hymenial layer well-developed, hyaline. All hyphae with clamp-connections.

Distribution: Sierra Leone, Uganda, Nigeria and other tropical countries.

MARASMIUS STRIGIPES Beeli (Fig. 13.)

This is a delicate species, usually growing solitarily, or in groups, on debris in shaded localities. Cap 7–17 mm in diameter, smooth, bell-shaped then expanded convex, obtuse, or subumbonate forming a papilla on the top of the pileus, dark brown to auburn-black, at least three quarters of the pileal surface is radially sulcate. Stem 3–7 cm long and about 1 cm thick, cylindrical, hollow, red-brown or concolorous with the cap, shining, smooth, white towards the apex; basal mycelium well developed, white to cream, strigose. Gills crowded to moderately crowded, narrow, free to adnexed, without interveining, white, often cream, edge whitish or concolorous. Spore-print white; spores hyaline, 12–16 × 3–4 μm, fusiform, smooth, inamyloid; cystidia ventricose to cylindrical, attenuate to appendiculate at the apex, thin or moderately thick-walled,

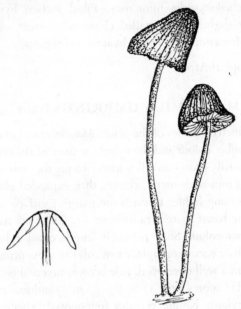

Fig. 13

generally not much differentiated from the basidia; cheilocystidia generally hyaline to subhyaline. Pileal surface hymeniform or consisting of cellular hyphae, individual hyphae hyaline, or golden coloured, basal part 11–18 × 4·9 μm, with the upper region covered by numerous setulae (3·5–8 × 0·7 × 1·4 μm). Context thin, consisting of thin-walled, hyaline and pseudoamyloid hyphae.

Distribution: tropics.

MARASMIUS LEVEILLIANUS (Berk.) Patouillard

This small, common species of *Marasmius* has a pan-tropical distribution. It mostly grows on dead leaves and twigs and on logs of wood, and are rust-brown in colour. The description given here is based on specimens found growing under bamboo trees, on fallen and decaying bamboo twigs, on the University of Ife campus.

Cap 10–20 mm in diameter, convex, becoming expanded, blood-red to deep purple; gills free, swelling and notched near the stalk, moderately spaced with often strong interveining. Stalk dark-brown, with a smooth, 'horny' and shiny surface, hollow, composed of three distinct layers: a surface layer of fairly thin-walled hyphae is heavily coated by a dark, resinous incrustation, forming a waterproof cuticle; a middle layer of very thin-walled, closely compacted, parallel hyphae; and an inner layer of thin-walled, filamentous hyphae which form the lining to the central cavity of the stalk. Spores 7–9 × 3–4 μm, oblong, slightly depressed on the adaxial side towards the apiculus, hyaline, thin-walled, contain one or more small oil-guttules, inamyloid; spore-print pure white. Basidia claviform, four-spored. Cheilocystidia numerous, thin-walled; pleurocystidia absent. Hymenophoral trama subregular, hyaline, thin-walled, with loosely interwoven hyphae. Subhymenial layer broad, subcellular. All hyphae provided with clamp-connections.

Distribution: throughout the tropics, mostly in Ceylon, Uganda, Kenya and Nigeria.

MARASMIUS JODOCODOS Hennings

Commonly grows among forest litter. Cap 1·5–4·5 cm in diameter, a deep violet in the centre and a pale lilac at the margin, smooth, dry, and glabrous. The gills are usually free, swelling in the middle, pale lilac in colour, slightly crowded and interveined. Stalk 3–10 cm in length, cylindrical, equal, hollow, yellowish-brown or light violet, glabrous, longitudinally striate, sometimes twisted. Context thin, white, strongly dextrinoid. Hymenophoral trama regular, hyaline, dextrinoid, consisting of thin-walled, parallel, hyphae. Subhymenial layer well-developed, subcellular. All hyphae with clamp-connections. Cheilocystidia abundant, intermixed with basidia, clavate, hyaline, thin-walled; pleurocystidia numerous, cylindrical, fusiform, hyaline, thin-walled, basidia clavate, four-spored; spore-print pure white; spores 6–10 × 3–4 μm, cylindrical, hyaline, thin-walled and inamyloid.

Distribution: Uganda, Nigeria and Cameroon.

MARASMIUS HAEDINIFORMIS Singer

This is a delicate species, found growing in large numbers on fallen plant debris in tropical forests. The cap is 6–25 mm in diameter, smooth, glabrous, bell-shaped, white to pale buff, radially striate, the striae corresponding to the underlying lamellae; gills either extending as far as the stalk but not attached to it, or broadly attached to the stalk, very narrow, much reduced towards the margin, concolorous with the cap, distant; stipe 2–6 cm long, hollow, medium-brown to deep brown, becoming much paler at the apex; context thin, white, dextrinoid, with thin-walled hyphae; spores elongate, subfusiform, sometimes curved at the apiculus, hyaline, thin-walled, smooth, inamyloid, non-dextrinoid; spore-print pure white; basidia clavate, bearing four short sterigmata; cheilocystidia numerous, forming a sterile gill edge, piriform to clavate, with the upper region heavily ornamented with short, cylindrical verrucae; pleurocystidia absent; hymenophoral trama regular, hyaline, consisting of thin-walled hyphae; subhymenial layer well-developed, subcellular; all hyphae with clamp-connections.

Distribution: Sierra Leone, Congo and Nigeria.

MARASMIUS FAVOLOIDES Hennings (Fig. 14.)

This species usually grows in damp tropical forests on leaf litter. The cap is lilac grey, 1·5–3·0 cm in diameter, at first convex finally becoming expanded, very thin, smooth, strongly radially ridged, margin entire; gills adnate to decurrent, cream, distant but strongly connected by prominent interveining to give a net-like appearance, edge delicately toothed; stipe 2–7 cm long, 1–4 mm wide, equal or attenuated towards the base, cylindrical, hollow, white at the apex but becoming brownish towards the base, smooth, abundant white mycelium attached to the base of the stipe; context very thin, concolorous, inamyloid, dextrinoid; hymenophoral trama subregular, hyaline, consisting of somewhat interwoven, thin-walled hyphae, inamyloid but strongly dextrinoid; subhymenial layer well developed, hyaline; all hyphae provided with clamp-connections; cheilocystidia and pleurocystidia absent; basidia clavate, four-spored; spore-print pure white, spores about 5–6 × 3–4 μm, ellipsoid, hyaline, smooth, thin-walled and inamyloid.

Distribution: tropics, mostly in Africa.

MARASMIUS ARBORESCENS (Henn.) Beeli (Fig. 15.)

Marasmius arborescens is one of the best known fungi in the tropics, for it is frequently seen in the earlier months of the year in its favourite habitat on old and decaying stumps of trees. The individual fruit body is inconspicuous being small, but the fruit bodies compensate for their small dimensions by their manner of sympodial growth forming a beautiful cluster drawing the attention of even the most casual observer. Cap 1–2 cm in diameter, bell-shaped then convex or expanded, white, whitish or slightly yellow-orange, dark brown in the centre, smooth, finely striated, transparent, slightly grooved; stipe 5–17 ×0·2 cm,

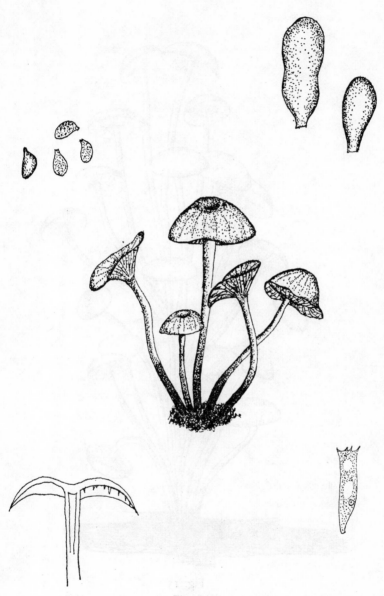

Fig. 14

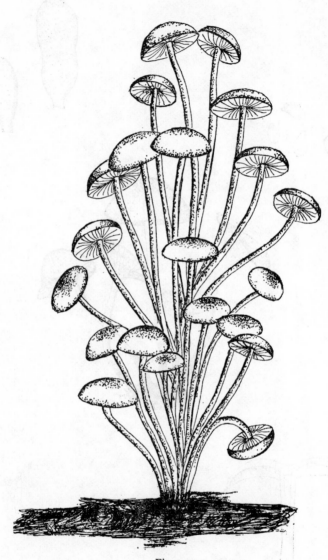

Fig. 15

equal or nearly equal, often compressed due to the pressure of the adjoining stipes, whitish above, reddish-brown or ochraceous-brown towards the base, densely fasciculate, joining each other near their bases; basal mycelium tomentose, cream-coloured; gills close, narrow, adnexed to slightly decurrent, pale yellow to cream, orange or whitish, with numerous lamellulae of various lengths, no interveining; flesh white or whitish, fragile; odour of bitter almonds, tastes sweet; spores hyaline, 9·2–10 × 3–3·5 μm, thin or moderately thick-walled, occasionally transforming into septate resting spores, inamyloid, smooth; cystidia absent; cheilocystidia present; pileal surface hymeniform; hyphae hyaline, claviform, smooth, thin or slightly thick-walled, often pedicellate, pseudo-amyloid; hymenophoral trama regular.

Distribution: mostly in tropics.

MARASMIUS GRANDISETULOSUS Singer (Fig. 16.)

This species grows on fallen twigs and dead tree trunks. Cap 2–2·5 cm in diameter, dark orange-brown in colour, at first campanulate, then campanulate-convex,

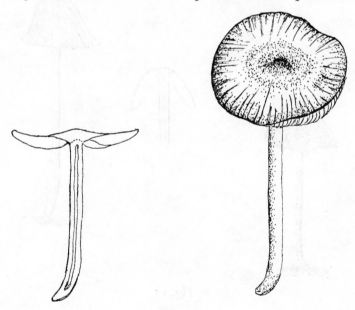

Fig. 16

naked, radially striated, margin slightly incurved, umbonate or slightly depressed in the centre, very distinctly furrowed, orange coloured grooves alternating with dark brown ridges; stipe 1·5–8·5 cm long and 0·5–2 mm thick, cylindrical, hollow, naked, smooth, at first white becoming amber towards the base and finally blackening with age; gills with or without lamellulae, sub-distant without interveinings, large to fairly large, cream coloured often with a distinctive pale orange-brown edge though sometimes concolorous, free or nearly so; flesh thin and whitish; odour spicy, taste acrid; spores hyaline, 15–21 × 3–4·5 μm, fusiform, somewhat curved, smooth, inamyloid; cystidia numerous, ventricose, rarely claviform, prominent; cheilocystidia present; pileal surface hymeniform, composed of stubbly hairs with ochraceous brown bases, 9–14 × 5–13·5 μm; pileal trama consisting of hyaline, pseudoamyloid hyphae.

Distribution: mostly tropics.

MARASMIUS GOOSSENSIAE Beeli (Fig. 17.)

This white *Marasmius* is a common species in many tropical countries and usually grows on marshy forest soils. Cap 2·5–7·0 cm in diameter, expanding to convex, recurved at the margin, umbonate or subumbonate or slightly depressed in the centre, slightly furrowed particularly towards the margin, cream or dull brown,

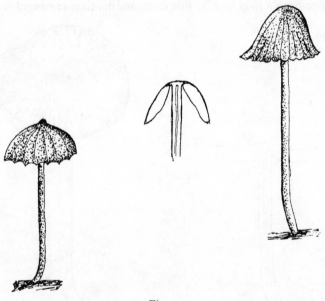

Fig. 17

darker in the centre; stipe smooth, 5–7 cm long, 3–6 mm thick, cylindrical, upper region concolorous with the cap, darker below, often bulbous at the base, attached to a strand of basal mycelium; gills moderately distant, often slightly interveined, free or nearly so, large, of a similar colour to cap margin; flesh whitish, thin, concolorous with the cap; spore-print white; spores hyaline,

6–7·5 × 3–3·5 μm, elliptoid or rarely ovoid, thin, inamyloid; cystidia many, present at the surface of the gills, thick-walled, hyaline, inamyloid, claviform; cheilocystidia present; pileal surface hymeniform or subhymeniform; hyphae smooth, globose to claviform, often pedicellate, hyaline; hymenophoral trama regular, pseudoamyloid, consisting of inflated thin-walled hyphae.

Distribution: mostly tropics.

CRINIPELLIS Patouillard

The species of this genus are very poorly known and occur on a number of substrata: on dead or living plants, usually on stems, roots, dead or living branches and fruits. The fruit bodies are covered with thick-walled hyphae which are distinctly hair-like and pseudoamyloid to almost amyloid, smooth, well separated from the trama, more rarely these hyphae are discoid with a number of appendages. Cap well-developed, gilled; cheilocystidia present; cystidia on the sides of the gills; spore-print white or whitish, spores hyaline, smooth, non-amyloid, thin-walled, but often becoming somewhat thick-walled, and very rarely (in only one species) septate; basidia usually four-spored, often more or less deformed; stipe central or eccentric, sometimes attached to white and occasionally pilose rhizomorphs; trama non-amyloid; hyphae with clamp-connections.

Distribution: cosmopolitan.

Practical importance: mostly parasitic. One species *C. perniciosa* has been reported to be the cause of the witch-broom disease of cocoa.

CRINIPELLIS DUSENII (P. Henn.) Singer (Fig. 18.)

A beautiful small fungus usually found growing on leaf-litter. Cap about 10 mm in diameter, conical bell-shaped becoming expanded, deep brown to vinaceous in colour; surface of the pileus covered with small hairs, 250–1300 μm in length, hyaline, changing to greenish-grey in KOH, strongly dextrinoid, simple, although occasionally lateral branches occur either near the apex or at the base; gills free, white, narrow and more or less crowded; stipe 2–4 cm in length, brown, darker towards the base, hairs of stalk similar to those of the pileus but shorter; context very thin, white, inamyloid, consisting of thin-walled hyphae; spores 6·5–9·5 × 4–5 μm, ellipsoid, hyaline, thin-walled, smooth, inamyloid, non-dextrinoid; basidia clavate bearing four sterigmata; cheilocystidia crowded, forming a sterile gill edge; pleurocystidia present, hyaline, scarcely projecting beyond

4

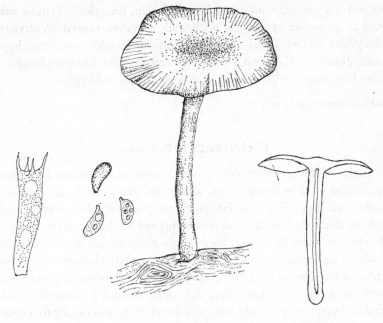

Fig. 18

the hymenium; hymenophoral trama regular or subregular, hyaline, usually composed of thin-walled hyphae; subhymenial layer subcellular; all hyphae provided with clamp-connections.

Distribution: tropics; common in Ghana and Nigeria.

FILOBOLETUS Hennings

The species belonging to the genus *Filoboletus* grow on dead wood, more rarely on other dead plant debris, and can be easily recognised by their truly poroid hymenophore, the pores form a rather deep or very shallow layer. The cap is convex; spore-print white or whitish; spores hyaline, smooth, ellipsoid, sometimes subglobose, amyloid; basidia normal; cheilocystidia not always clearly differentiated from the pseudoparaphyses; stipe central or slightly eccentric subglabrous to pruinose or flaccose, not tomentose except occasionally at the base; context of thick-walled hyphae, very frequently somewhat gelatinised, non-amyloid, with clamp-connections.

Distribution: tropics; Japan, Florida and Nigeria.

Practical importance: none.

FILOBOLETUS GRACILIS (Klotzoch ex Berk) Singer

This poroid mushroom is not very common in the tropics. It usually grows on dead wood in a moist and warm climate. In spite of its poroid hymenophore, the fungus obviously belongs to the order Agaricales. Cap 4 cm in diameter, convex with a small pointed umbo, smooth, russet colour; flesh thin, concolorous, soft and flexible; hymenium white, poroid; pores minute, about 250 μm across and not exceeding 500 μm in length; stipe slender, smooth, cartilaginous, hollow, concolorous; spore-print white; spores globose, 5·5–6 μm, hyaline, amyloid; basidia normal, four-spored, 15 × 15 μm; cheilocystidia thin-walled, 20 × 7 μm, vesiculose; hymenophoral trama consists of loosely-woven hyphae, with rather thick hyaline walls, embedded in a gelatinous matrix; pileal surface with non-gelatinised more compact hyphae.

Distribution: tropics and sub-tropics.

Practical importance: none.

XEROMPHALINA Kühn. & Maire

The species of *Xeromphalina* are rare tropical fungi. They usually grow on dead or living parts of the plants, on buried wood or on humus, and are recognised by the bright colours of their carpophores. The cap, with incurved margin, possesses an epicutis which consists of radial, repent, non-diverticulate hyphae; hymenophore lamellate but in some species it becomes sinuate, intervenose; gills broadly adnate to deeply decurrent, coloured; hymenophoral trama regular; basidia normal; cystidia present on the edge or frequently on the side of the gills, hyaline; stipe more or less central, coloured and with a tomentose base, without veil; context tough, non-amyloid; hyphae with clamp-connections.

Distribution: common in temperate zones, rare in tropics.

Practical importance: parasitic on stumps and trunks of living plants.

XEROMPHALINA TENUIPES (Schwein.) A. H. Smith (Fig. 19.)

This beautiful species is very widely distributed in tropical and temperate regions. It is an extremely variable fungus in its macroscopic and microscopic characters and has been described many times under different names. Cap 1–8 cm diameter, at first convex then becoming expanded and broadly umbonate, brownish in the centre, becoming yellowish towards the margin, occasionally entirely yellow, striated towards the margin, often reticulately rugose, pruinose to velutinous;

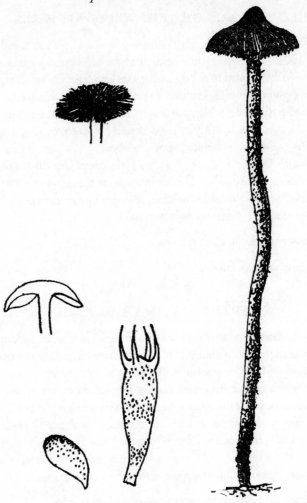

Fig. 19

margin paler, thin, entire; gills whitish, sinuate, adnexed, usually with a decurrent tooth, slightly ventricose, moderately distant, with various sized lamellules
and conspicuous interveinings, edge entire, slightly pruinose; stipe up to 8 cm
long, tough and flexuous, cylindrical and hollow, equal or slightly swollen towards
the base, occasionally with a small, tapering pseudorhiza, surface velvety, concolorous towards the apex, darker towards the base; context brown, inamyloid,
consisting of densely interwoven, hyaline hyphae with very thick gelatinised
walls; spore-print pure white; spores 6–9·5 × 3–5 (average of 7·5 × 4) μm, ellipsoid to cylindrical, flattened or slightly concave, hyaline, amyloid, smooth,
thin-walled, with one or two oil-guttules; basidia clavate, hyaline, four-spored;

cheilocystidia filiform, forming a sterile narrow gill-edge, hyaline, thin-walled; cystidioles abundant, fusiform, hyaline, thin-walled. Hymenophoral trama subregular to irregular, hyaline, inamyloid, consisting of loosely interwoven thick-walled hyphae; pileal surface has an epicutis; all hyphae with clamp-connections.

Distribution: tropical and temperate regions; reported from India, Pakistan, Australia, Congo, Kenya, Nigeria, South Africa, U.S.A., Trinidad and Brazil.

FAVOLASCHIA (Pat.) Hennings

The species of the genus *Favolaschia* commonly grow on dead wood in the tropics. Generally characterised by the poroid, alveolate, sublamellose, anastomosing or venose – anastomosing hymenophores, with more or less gelatinous trama; pileus eccentrically or laterally stipitate, or sessile to resupinate with a definite margin, small, usually diaphanous; hymenium inferior, consisting of alveolate tubes, or veins, or obtusely anastomosing lamellae with or without connecting veins; cuticle of the pileus is either undifferentiated or consists of a thin layer of diverticulate hyphae; trama gelatinous; hyphae non-amyloid, clamped; hyphae of the hymenophoral trama interwoven or very loosely arranged and wavy-subirregular in a gelatinous mass; spores medium to rather large, 5–16 μm, broadly cylindrical or more often ellipsoid, or subglobose, hyaline, smooth, amyloid, thin-walled; basidia unicellular; occasionally long, often short and thick, two- or four-spored; sterile bodies of the cuticular layer sometimes enter the hymenium which is continuous with the hymenophore; cystidia, gloeocystidia, dendro-physes and pseudophyses frequently present.

Distribution: temperate, tropical and sub-tropical zones.

FAVOLASCHIA THWAITESII (Berk. & Br.) Singer

The small and gregarious fruit bodies of this species frequently appear on the dead logs of wood in forests. The species can be recognised by its small, orange to orange-red laterally stipitate fruit body and wide polygonal or round hymenial pores. Cap 5–10 mm wide, reniform, plano-convex, orange-coloured, translu-cent, margin entire; hymenophore poroid, concolorous with the upper surface of the cap; pores polygonal, 1–2 per mm, separated by moderately thin dissepi-ments, pores wider and more angular towards the stipe, narrower towards the margin; stipe 2–5 mm long and lateral, concolorous with the cap, translucent, arising from a mycelial base; context thin, hyaline, consisting of freely branching, inamyloid hyphae; spore-print white; spores 7–13 × 4–8 μm, ellipsoid, hyaline, amyloid, smooth, thin-walled; basidia clavate, cylindrical, four-spored; cheilo-cystidia abundant, varying in shape, hyaline, thin-walled; pleurocystidia absent; gloeocystidia cylindrical, clavate, thin-walled; hymenophoral trama subregular, hyaline, inamyloid, consisting of widely spread, thin-walled hyphae. Pileal surface well differentiated, forming an irregular epicutis; all hyphae with clamp-connections.

Distribution: tropics; frequently found in Uganda, Kenya, Zambia and Nigeria.

AMANITACEAE

TERMITOMYCES Heim

This genus is distinguished from all other genera of agarics by the distinctive pseudorhiza and its obligate association with termite nests. The cap of the fruit body has a prominent umbo; the cuticle consists of repent, filamentous and hyaline hyphae; the gills are usually free to subadnate, either having a notch at the point of attachment to the stipe, or have a decurrent tooth; hymenophoral trama initially bilateral, then becoming regular – subintermixed; cystidia present; basidia normal; spores smooth, ellipsoid, hyaline, with continuous homogeneous, rather thin-walled; spore-print pink; stipe with a long and tapering pseudorhiza reaching the termite nest some distance below the soil surface. (The genus obtains its nourishment from the material of the termite nest which is constructed of woody material masticated by the termite workers). Context compact, fleshy, or somewhat tough in the stipe; hyphae without clamp connections. The primordia usually develop in the holes of the termite nests and can be observed by removing the soil from where the carpophore is growing and exposing the nest.

Distribution: cosmopolitan. Several species of *Termitomyces* are now well-known developmentally and biochemically. Commonly occur in the tropical regions of Asia, Africa and south Pacific.

Practical importance: most species are highly valued as foods and are considered by some to be superior to all other mushrooms.

TERMITOMYCES STRIATUS (Beeli) Heim (Fig. 20.)

This species grows either on or near termite hills and is a medium-sized fungus with a cap up to 7 cm in diameter, convex with upturned margin, brown in colour, darker in the centre with an indistinct umbo; gills free, crowded, orange coloured; stipe from 2·8 to 7·2 cm in length, concolorous; pseudorhiza up to 4·5 cm deep, tapers gradually towards the farthest end; flesh pale grey; cystidia numerous, piriform, clavate, or cylindrical; basidia clavate, two- to four-spored; spores smooth, hyaline, thin walled, ellipsoid (16–21 $\times$ 4–5 μm); sporeprint pinkish-cream.

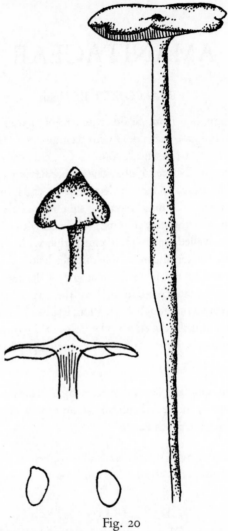

Fig. 20

Distribution: widespread in tropics, common in Ivory Coast, Cameroon, Congo, Sierra Leone, Guinea and Nigeria.

Practical importance: edible.

TERMITOMYCES ROBUSTUS (Beeli) Heim (Fig. 21.)

The species is recognised by the large carpophore. Cap dark tawny-brown, scrobiculate, 6–18 cm in diameter, margin inflexed; gills free, white, crowded, stem equal, greyish-brown, solid, 4·5–6·0 cm in length, continues below the soil

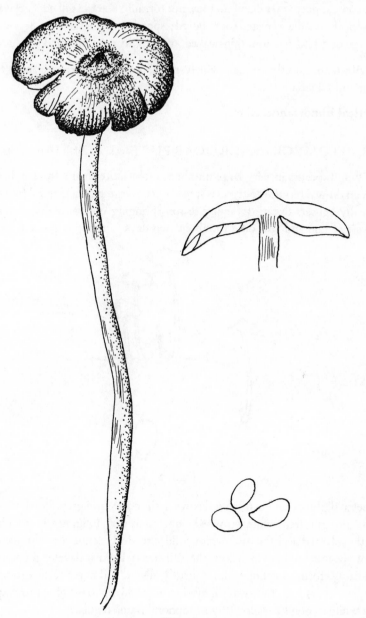

Fig. 21

surface as a 17·3–28·0 cm long, equal, pseudorhiza which attaches to the termite nest; context grey; cystidia numerous, not forming a sterile gill edge, clavate or cylindrical; basidia clavate, four-spored; spores fairly large, 5·7–8·5 × 4–5·5 (average of 7·5 × 4·8) μm, thin-walled, smooth, hyaline, ellipsoid.

Distribution: mostly tropics, widely distributed in Guinea, Uganda, Sierra Leone and Nigeria.

Practical importance: edible.

TERMITOMYCES MICROCARPUS (Berk & Br.) Heim (Fig. 22.)

This fungus occurs in very large numbers, often more than a thousand at one time, on disused termite nests and is spread over a large area of tropical forest soil. It usually appears during the rainy season springing up overnight after a few showers of rain and lasts only for about two days. The cap is about 1–2 cm in

Fig. 22

diameter, slightly umbonate, olive-brown in the centre and greyish-yellow along its margin; gills free, white, crowded; stipe up to 5 cm long, nearly straight but slightly enlarged near the base, white, solid; pseudorhiza not very distinct, often absent, penetrates 1–3 cm below the soil level; basidia clavate, four-spored; cystidia pyriform to cylindrical, frequently absent, and never form a sterile gill edge; spores smooth, thin-walled, hyaline, ellipsoid, up to 6–7 × 3–4 μm; spore-print white to greyish-yellow; hymenophoral trama regular.

Distribution: in the tropics; Ceylon, Sierra Leone, Nigeria, Kenya and Tanzania.

Practical importance: edible.

TERMITOMYCES GLOBULUS Heim & Goossens (Fig. 23.)

The fruit bodies emerge during the rainy season on the moist soil. They are found in association with termite nests. Cap dull-orange coloured, grows up to about 10 cm in diameter, and finally becomes convex with a turned-up margin; gills joined to the stipe having a notch at the point of their attachment, white,

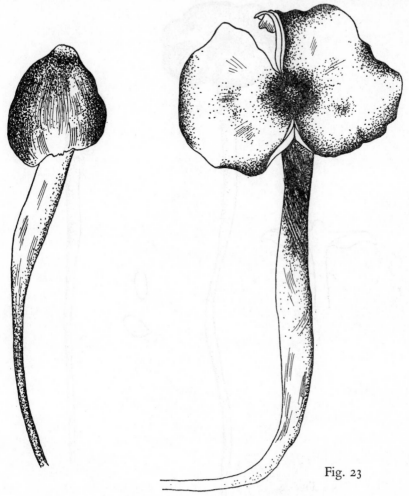

Fig. 23

crowded; stipe up to 8 cm in length and 1·2 cm in diameter, usually equal, white, smooth and solid; pseudorhiza up to 11 cm long, equal, rarely bulbous, solid; flesh usually pale grey; basidia clavate and four-spored; spores smooth, thin-walled, white, ellipsoid and 5 × 4 μm.

Distribution: almost all tropical regions of the world.

Practical importance: edible.

TERMITOMYCES MAMMIFORMIS Heim (Fig. 24.)

This is a very common species which usually grows either in groups on termite hills or occasionally singly on the soil in association with termite nests. The species can be identified by the presence of a persistent annulus. The cap is usually 7 cm in diameter with a very prominent umbo and inflexed margin; gills free, crowded,

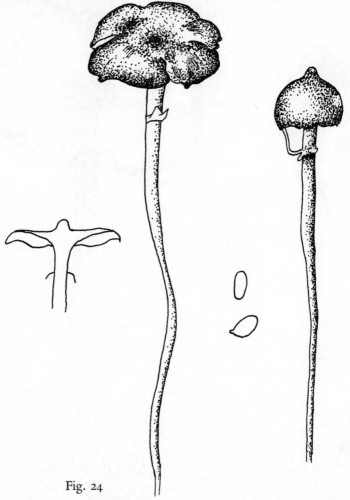

Fig. 24

white; stipe up to 5 cm in length and 1 cm in width, equal, white, solid; pseudo-rhiza up to 9 cm deep, solid, tapering gradually towards the lower end which is attached to the termite nest; flesh white; cystidia clavate – pyriform; basidia clavate; spores smooth, hyaline, ellipsoid and 6·3 × 3·3 μm in size; spore-print greyish-yellow.

Distribution: almost all tropical zones.

Practical importance: edible.

TERMITOMYCES CLYPEATUS Heim (Fig. 25.)

Frequently grows on shaded forest soil in association with termite nests. It is distinguished by its characteristic silky, greyish-brown cap and with reflexed margin at maturity, rarely exceeding 7 cm in diameter, and bearing a smooth, dark-coloured, strongly spiniform, umbo; gills free, white, crowded;

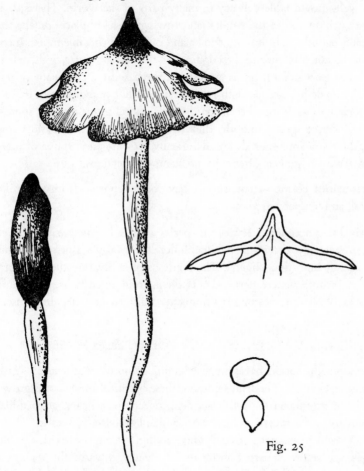

Fig. 25

stipe up to 8 cm long, usually equal but rarely becoming bulbous near the soil surface, white, solid, and continues below the soil surface as a long pseudorhiza, 2·5–8 cm in length; flesh white; cystidia clavate to turbinate, numerous; basidia clavate two- or four-spored; spores smooth, thin-walled, ellipsoid, hyaline 5–8 × 3–4·5 μm; spore-print pinkish.

Distribution: tropics; mostly in Ghana, Kenya, Zaire and Nigeria.

Practical importance: edible.

VOLVARIELLA Spegazzini

The young and healthy fruit bodies of some of the species of the genus *Volvariella* are a well-known table delicacy in many parts of the world. There are many species encountered in the tropics, all growing in shady places on the soil, on decaying palm logs, in hollow trunks and on other vegetable refuse. It appears during the rainy season and is recognised by its pink spores free gills, and a stipe which bears no annulus but is enclosed at the base by a cup-shaped persistent 'volva'. Cap fleshy, circular with a central stipe, white or pigmented; gills always free, forming a ring round the stipe; hymenophoral trama inverse; spore-print brownish-pink; spores smooth, thick-walled, non-amyloid; basidia normal, two-, three-, or four-spored; cystidia usually present; pileal trama of branched and interwoven hyphae, often with no clamp-connections.

Distribution: cosmopolitan; species have been reported from tropical, subtropical, and temperate regions.

Practical importance: a few of the species, such as *V. volvacea* and *V. diplacea*, are commercially grown in China, Indonesia, Malaya, Burma, Philippines, Madagascar, India and Nigeria. The methods of cultivating these mushrooms are often primitive and differing due to the use of different substrata in different countries. In Nigeria, the fungus is frequently cultivated on oil palm refuse.

VOLVARIELLA VOLVACEA (Bull. ex Fr.) Singer

In the tropics this species has as much economic importance as *Agaricus bisporus* in the temperate zones. The fruit bodies of these beautiful mushrooms grow solitarily or in groups on the soil in gardens, on compost heaps, at roadsides, or occasionally on the mortar of brickwork and in hothouses and cellars. The cap is from 5–10 cm in diameter, ovoid when young but expands with age, fleshy, thick in the centre, tapering more or less evenly towards the margin, dry, fuliginous to greyish-brown. Stem 4·5 to 14 cm in length, off-white to dull-brown in colour, base slightly bulbous, volva large, with a free margin and irregularly lobed, membranous, brownish; gills close, broad, free, white at first but finally becoming deep flesh-coloured; spores 7–10 × 4–7 μm, ovoid; basidia clavate, four-spored; pleurocystidia usually sub-cylindrical, abundant; cheilo-cystidia usually fusoid – ventricose or clavate, abundant; hyphae without clamp-connections.

Distribution: mostly in tropics.

Practical importance: edible; cultivated commercially in several tropical countries.

VOLVARIELLA SPECIOSA (Fr. ex Fr.) Singer

This white mushroom is quite common and grows on richly manured ground on lawns, gardens, fields and woods. The cap is smooth and very viscid when moist, generally white, occasionally pink, bell-shaped when young but finally becom-ing convex to expand, from 5–15 cm in diameter; stipe firm, tough, and attached to the centre of the cap, solid, bulbous at the base, more or less concolorous with the cap, smooth, dry, without an annulus and with a membranous white volva which is persistent and remains attached to the slightly bulbous base of the stipe, gills brittle, easily separated from the pileus, extending as far as the stem but not attached to it, broad, thick and of different lengths, white at first, when the spores are immature, but finally become deep salmon-coloured; cheilocystidia numer-ous, fusoid, or of various other shapes; pleurocystidia sub-cylindrical and fuscous, clamp-connections absent; basidia clavate, four-spored; spores smooth, deep-salmon, ovoid occasionally obovoid, about 9–18 × 6–10 μm in size.

Distribution: cosmopolitan.

Practical importance: edible.

VOLVARIELLA BOMBYCINA (Schaeff. ex Fr.) Singer (Fig. 26.)

This is a rather large species, and usually grows during the rainy season on rotting wood, leaf litter and in richly manured soil, particularly in the palm and coffee plantations. Cap soft, fleshy, generally white, yellowish at first, then white be-coming fawn, 5–20 cm in diameter, globose to ovoid then expanding, dry, silky fibrillose becoming squamulose, with the margin more or less fimbriate; flesh thin, soft, white, with a mild taste and a slight odour; gills crowded, broad, ventricose, free, white, becoming flesh-coloured; stipe up to 20 cm long and 2 cm thick, tapering towards the apex, without an annulus, with a bulbous base which is encased in a large volva; spores 7–9 × 5–6 μm usually ovoid, occasion-ally oval, oblong, or obovoid; basidia clavate, four-spored; pleurocystidia

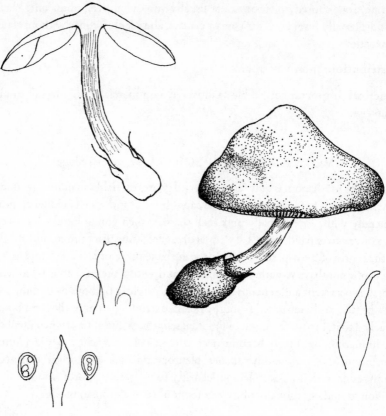

Fig. 26

usually fusoid, fairly to very abundant; cheilocystidia fusoid, often of various other shapes, abundant; hyphae without clamp connections.

Distribution: cosmopolitan.

VOLVARIELLA ESCULENTA (Mass). Fries (Fig. 27.)

A common tropical edible mushroom it usually grows abundantly during the rainy season and frequently appears in the hollow trunks of rotting oil-palm trees. It is rather fascinating to watch the outer skin of the fruit body crack and the cap emerge leaving the skin as a volva at the base of the stipe. Pileus at first whitish towards the margin and yellowish-brown in the centre and entirely punctate – velutinous, the punctations being violet-black or brown, the margin remaining light coloured, even whitish or shining white, often with long and fine radial streaks which are silvery grey, becoming more and more dense towards the centre, brownish in the zone surrounding the umbo; surface smooth,

Fig. 27

at first rugose and punctate concentrically, then very finely remulose; margin at first broadly involute, irregular and unequally appendiculate with patches of the universal veil, later fimbriate, globose, then piriform, soon becoming ovoid – cylindrical; pileus at maturity becomes campanulate and umbonate and eventually when it is fully open it becomes convex, reaching 14 cm in diameter; gills reddish-pink, narrowed towards the stipe, close, thin, free; stipe whitish, with fine longitudinal cream coloured streaks, solid, 8–12 cm long, bulbous at the base; volva membranous, cylindrical, thick, tearing off irregularly, persistent, violet-black-grey to greyish-brown, minutely fibrillate; context brown under the cuticle, white and soft inside the pileus and volva, cream-coloured and fibrous in the stipe; spore-print salmon pink; spores 9–15 × 4–7 μm, smooth, thin-walled; basidia clavate, four-spored; cheilocystidia abundant, broadly clavate, hyaline; pleurocystidia scattered; hymenophoral trama inverse.

Distribution: mostly tropics.

Practical importance: edible.

AGARICACEAE

CHLOROPHYLLUM Massee

The fruit bodies of this genus are frequently encountered growing on rich soil in most tropical countries. Cap scaly, epicutis consisting of a palisade of erect hyphae on the disc; gills usually becoming green, quite free and remote from the stipe; spore-print of various shades of green, or more rarely colonial buff; spores smooth with thick, complex wall which stains with cresyl blue, with a broad germ pore, large; cystidia absent; cheilocystidia present; hymenophoral trama almost regular; stipe long, bulbous at the base, but without a volva, the upper part of the stipe has a movable annulus which is tightly attached to the stem in young and fresh specimens but becomes free on drying; context becomes reddish on bruising, containing a poison; hyphae non-amyloid, without clamp-connections.

Distribution: mostly tropics. Common in tropical America, Oceania, Asia, North America, South America and Africa.

Practical importance: most of the species are *poisonous*. Some are reported to be edible, but this indication must be regarded with caution.

CHLOROPHYLLUM MOLYBDITIS (Mayer ex Fr.) Massee (Fig. 28.)

The magnificent fruit bodies of this mushroom are a familiar sight on rich forest soil and at the roadsides. The species can be recognised by its large scaly cap and unique greenish colour of the gills and spores. Cap 5–24 cm in diameter, soft and fleshy, at first globose then expanding to plane with a broad umbo, cuticle brownish, breaking up into scales which are concentrated near the centre of the cap, the rest of the cap is ornamented with minute deciduous silk-like fibres; margin entire, striate for a short distance. Gills free, pale pinkish-buff, then becoming greenish, often staining reddish when bruised, ventricose, fragile, moderately distant, with lamellulae of various lengths; stipe up to 28 cm long, easily separable from the pileus, bulbous base, cylindrical, fibrous, annulus attached at the upper part of the stipe, movable when dried, thick, fleshy; context of pileus pale pinkish-cinnamon, reddening on bruising, inamyloid, consisting of interwoven thin-walled septate hyphae; context of stipe tough-fibrous; spore-print greenish, fading on drying; spores 8–11 × 6–8 μm, obovoid to broadly

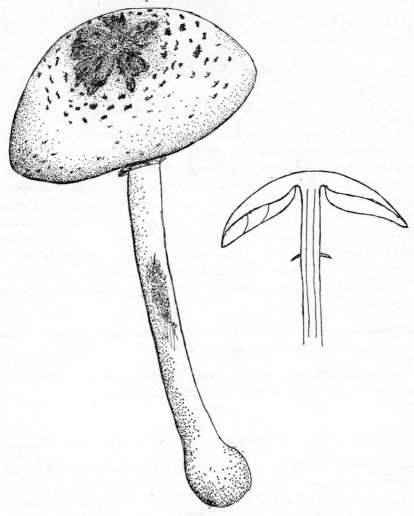

Fig. 28

ellipsoid, smooth, greenish, dextrinoid; basidia clavate, four-spored; cheilo-cystidia abundant, hyaline, thin-walled; pleurocystidia absent; hymenophoral trama narrow, regular, hyaline, inamyloid, consisting of thin-walled hyphae; pileal surface consists of erect, septate hyphae with thickened-walls; all hyphae without clamp-connections.

Distribution: widely distributed in the tropical and subtropical regions; reported from Australia, Canada, U.S.A. and Africa.

Practical importance: there is a great deal of confusion about the nature of this species, some forms are reported as edible while others are regarded as poisonous;

it has been suggested that the toxicity is dependent upon climatic and habitat factors.

LEUCOCOPRINUS Patouillard

This is a common tropical genus which usually grows either on shaded soil, or on lawns, and on various other substrata. The marginal part of the cap usually splits at maturity, and is sometimes covered with spherocysts; gills free and often form a collarium round the stipe, thin and soft; spore-print pure white to yellowish; spores with more or less distinct germ pore, distinctly metachromatically coloured in cresyl blue; cystidia either rare or absent; cheilocystidia numerous; clamp-connections absent; context thin, non-amyloid; hymenophoral trama more or less regular; stipe often sub-filamentous and long and usually with a movable annulus; volva absent.

Distribution: tropics.

Practical importance: a single yellow species of the genus has been reported to be very poisonous.

LEUCOCOPRINUS CEPAESTIPES (Sow. ex Fr.) Patouillard (Fig. 29.)

This shining white mushroom is very common in the tropics and appears, often

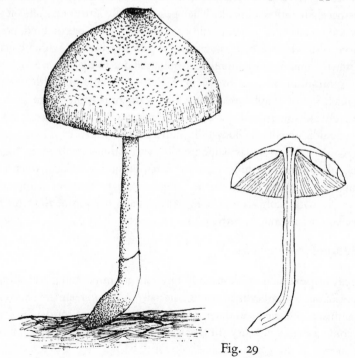

Fig. 29

gregariously, during the wet season on the ground in lawns and among grasses, and occasionally on the dead logs of wood. Cap up to 8 cm diameter, at first ovoid, then conical with a rounded umbo, becoming campanulate or expanded, plicate-striate at the margin only, pure white, later becoming cream or buff at the disc only, covered with small, loose mealy scales; flesh soft, moderately thick, white; gills crowded, thin, broad, free, white; stipe with a fusiform base 6–8 mm thick, equal and 2–3 mm thick above, hollow, white or cream coloured; ring membranous, movable, white, soon breaking up, stipe mealy below the ring. Spores ovoid with a germ pore, red-brown with Melzer's reagent, size variable, mostly 8–11 × 5–7 μm.

Distribution: tropical.

AGARICUS L. ex Fries

Most of these mushrooms are a well-known table delicacy in many parts of the world, and can be recognised by their spore colour and the lack of spherocysts in the cortical layers. However, due to their enormous variety the determination of the species is rather difficult. The species usually grow on soil, on dung, humus and on ant hills, in and outside woods. Cap white or coloured, naked or squamose, also with warts, or smooth dry; gills free but not with a collarium, white when young but becoming deep coloured due to the presence of attached spores; spore-print purplish-brown to sepia; spores smooth, with compound wall which is not visibly pseudoamyloid, with usually no germ pore, very occasionally indistinct; basidia normal in all regards, but often consistently two-spored; cystidia absent; cheilocystidia numerous in some species; hymenophoral trama regular then irregular; stipe usually with a thin-membranous fugacious to thick, almost fleshy annulus; context often becoming reddish when bruised or changing to yellow when touched; hyphae usually without clamp-connections, rarely with clamp-connections; surfaces and context of the fruit bodies occasionally react strongly with ordinary reagents.

Distribution: cosmopolitan.

Practical importance: economically this is a very important genus, all species, but a poisonous few, are edible. Some, especially *A. bisporus*, are grown commercially in many countries, and the production of carpophores for the food market has become a major industry. In tropical countries *A. bisporus* is replaced by edible species of the genus *Volvariella* (*q.v.*).

AGARICUS TRISULPHURATUS Berkeley

This pink-coloured species generally grows on sandy soils either in gardens or forests and is identified by its bright colour and copious woolly veil which can be easily detached. Cap 3–10 cm in diameter, fairly thick, globose-campanulate then convex, finally expanded or slightly depressed, generally regular, occasionally umbonate, margin becoming incurved with age; the scales are bright orange, disappearing with age, the surface becoming ochraceous – pink and orange; margin for a long time remains attached via the veil at least during the young stage; stipe 3–6·5 cm long and 0·5–1·3 cm thick, cylindrical, stiff, rarely flexible, hollow, sometimes radial, yellowish and villose at the top, the lower part is covered with bright-orange pulverulent flakes of the veil; annulus fluffy, fragile, not persistent; gills large, edged with teeth, thin, free, narrow, sharp pointed at the margin, ventricose, white-pink then deep brown; edge brownish; context fleshy firm, fibrous, in the stipe, pinkish white, then brownish, sometimes tinted with bright yellow in the upper part of the stipe; spores dark brown 4·6–6·5 × 3·2–4·5 μm, ellipsoid or ovoid; cheilocystidia abundant, claviform generally with a narrow base, rarely ellipsoid or subglobose, and with a thin yellow membrane, edges more or less incrusted with extracellular yellow pigment. The pileal surface covered with a thick-layer of declining hyphal elements, cylindrical, sometimes branched, 4–9 μm in diameter, orange with a thin yellow membrane, smooth or having fine warts. The surface of the stipe below the annulus is of the same colour as the cap. Reaction with guaiac tincture grey-green then darkish, rapid; with guaiacol fairly rapid, changing to purple; with ferrous sulphate negative; Schaffer's cross-reaction (aniline × HNO_3) nil.

Distribution: tropical Africa and India.

AGARICUS MURINACEUS (Beeli) Pegler

The little fruit body of this species is fairly rare in the tropics, growing on sandy soils. Cap 2–3 cm in diameter, convex then expanded, broadly umbonate, surface pale grey, covered by small, sepia-brown, sub-erect scales, margin striate, undulate; gills more or less free, broad, dark sepia, crowded with lamellulae, edge white pruinose; stipe 2–3 cm long, equal, cylindrical, with a sub-bulbous base, hollow, smooth, white or pale grey; context thick and white; spores 5–6 × 3–4·5 μm, ovoid, smooth, thick-walled, without a germ pore; spore-print fuscous; basidia oblong, two- or four-spored; cheilocystidia numerous, ovoid, cylindrical, hyaline, thin-walled; pleurocystidia absent; hymenophoral trama regular or nearly so, subhymenial layer well-developed, sub-cellular; all hyphae devoid of clamp-connections.

Distribution: reported from Sierra Leone and Nigeria.

AGARICUS IODOLENS Heinem (Fig. 30.)

This species is characterised by its large size, brown or dark-brown cap covered with small fibres or scales and its peculiar iodoform smell. Cap fairly thick, 10–20 cm in diameter, when young enclosed within a universal veil, when expanded fragments of the veil remain attached to the margin of the cap; pileus

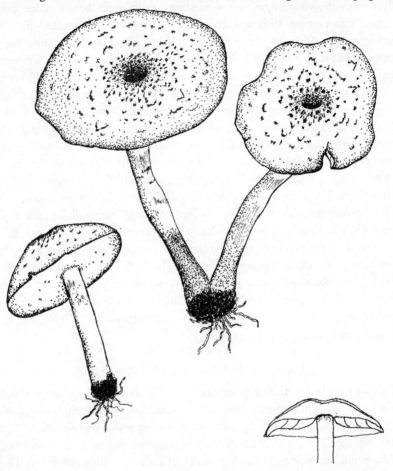

Fig. 30

globose at first then becoming bell-shaped, convex at the top, margin incurved, truncate; cuticle brown to dark-brown, darker in the centre, with a coating of small silky fibres and patches of scales; stipe up to 18 cm long and 1·2–2·5 cm thick, straight or curved near the base, not bulbous; cuticle smooth, silky white to dirty ochraceous, becoming light yellow when touched; annulus near the upper end, wide, thick, white, occasionally yellow; mycelial strands numerous;

gills close, thin, unequal, narrow, swelling in the middle, when young the gills are very white, gradually becoming pink, brown, and finally dark-brown at maturity; flesh firm, white, becoming chrome-yellow or deep orange, particularly at the base of the stipe, the yellow colour disappears rapidly, and leaves the flesh pink; spores purple-brown, 5–7 × 3–4 μm, ellipsoid or slightly ovoid; cheilocystidia claviform, membranous, thin, hyaline; taste sweet but disagreeable, odour of iodoform; reaction with ordinary guaiac tincture is positive; with gyaiacol, the colour changes rapidly to deep red or crimson-purple; with ferrous sulphate, no discoloration or negative; with phenol, wine colour, slow reaction; with alpha-naphthol, rapid reaction, turning deep red; with pyrogallol, the colour of the context slowly changes to brownish-yellow; and no reaction with phenol-aniline.

Distribution: tropical.

AGARICUS ALPHITOCHROUS Berk. & Broome

This pink-coloured mushroom grows in groups along curved lines on the ground and often makes its appearance on moist sandy soil under the shade during the early months of the rainy season. The entire surface of the fruiting body, including the part of stem below the annulus, is usually covered with minute, pink-coloured, hair-like structures. Cap 2–5 cm in diameter, convex then expanded plane, rarely umbonate, surface pale vinaceous, with pink-coloured minute scales; margin entire, thin; gills free, pale-salmon coloured, slightly ventricose, crowded, with lamellulae of various sizes, edge entire, concolorous; stipe up to 6 cm long, equal, often with a swollen base, hollow, cottony white; ring membranous, persistent, flesh thin with interwoven, hyaline, thin-walled hyphae; spores 5–7 × 3·5–5·4 μm, ovoid, thick-walled, without a germ pore; basidia clavate, four-spored; cheilocystidia abundant, sub-globose, piriform or clavate, hyaline, thin-walled; pleurocystidia absent; hymenophoral trama regular, hyaline, inamyloid, consisting of more or less parallel, thin-walled hyphae; pileal surface consisting of thin-walled, short, inflated interwoven elements; all hyphae without clamp-connections; with guaiac, pileal surface slowly turns blue; with phenol, red; with ferrous sulphate, pale green; Schaeffer's cross-reaction (aniline × HNO_3), nil.

Distribution: mostly tropics; reported from Ceylon, Ghana, Kenya, Tanzania and Nigeria.

LEPIOTA (Pers. ex Fr.) S. F. Gray

Many species of *Lepiota* are found in the tropics, where they usually grow on the soil and on various living or dead substrata. Cap often umbonate, covered with scales of a micaceous or fibrous texture, often forming a smooth disc in the centre of the cap, but the original entire cuticle breaks up all around the disc into fragments which are often deeper coloured than the surface of the cap; stipe detaches easily from the cap, and is long, straight or curved, hollow, often bulbous at the base; veil generally simple, membranous persistent, forming a ring around the stipe, which is sometimes movable around the stem when old. The hyphae of the epicutis very rarely repent and filamentous; trama usually non-amyloid; gills free, remote, separated from stem by a collar; cystidia sometimes present on the sides of the gills but absent in the majority of the species; cheilocystidia mostly present, elongated, conspicuous; basidia clavate, four-spored; spore-print white, often cream in colour, light-brown to burnt-amber; spores hyaline, ellipsoid or ovoid, with complex walls, pseudoamyloid or rarely amyloid, without distinct germ pore; hyphae with or without clamp-connections.

Distribution: cosmopolitan.

Practical importance: inedible; few species are poisonous, the symptoms are mostly phalloidic, and often the poison is fatal.

LEPIOTA BIORNATA (Berk. et Br.) Saccardo (Fig. 31.)

This ivory-white mushroom is quite common in the tropics and appears in tufts on dead wood or palm logs and soil. Cap hemispherical, becoming conico-campanulate and expanding to about 5 cm, surface vinaceous buff, almost concealed by crowded, minute, russet to chocolate brown scales; flesh pale yellowish to almost white, rapidly staining with xanthin orange on exposure to the air; gills crowded, broad, free, ivory yellow, the edges staining like the flesh; stipe attenuated upwards from about 5–7 mm to 4–5 mm, concolorous with the pileus and scurfy – squamulose below the ring, solid; ring large, membranous, sheathing the stipe, white above, concolorous with the pileus and squamulose below; spores broadly elliptical, dextrinoid, 7–9×5–6 μm; cheilocystidia thin-walled, clavate, 40–55×12–20 μm, some with narrow cylindrical beaks about 12×3 μm, contents granular; pileus surface covered with erect,

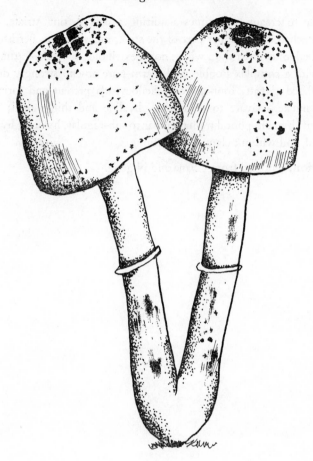

Fig. 31

ventricose, obtuse hairs up to 210×15 μm, dark brown at the base, shading to hyaline at the tip, forming a smooth disc 'calotte' which cracks into squamules.

Distribution: tropics; mostly in Trinidad, Spain, Ceylon and Nigeria.

LEPIOTA GOOSENSIAE Beeli

This deep red *Lepiota* is quite common in the tropics and grows on the soil, under shade, and in forest zones. The cap is about 2–5 cm in diameter, convex becoming expanded; colour of the cap is coral red, darker in the middle with fine radial striations which become disorganised in older specimens, the cap then becoming minutely squamulose. Gills free, white, narrow, edges white and denticulate; stipe 2 to 9 cm in length and easily separable from the cap, slightly bulbous at the base, white to cream, sometimes acquiring a pinkish tint, striate, glabrous; annulus attached to the upper part of the stalk, membranous, persistent, white with a red edge; context thin, white, non-dextrinoid; spores 6–8 μm, ellipsoid, hyaline, with a complex double wall, germ-pore absent, strongly dextrinoid; basidia inflated clavate, four-spored; cheilocystida present and form a sterile gill-edge, inflated-clavate to cylindrical, hyaline and thin-walled; pleurocystidia absent; hymenophoral trama regular to sub-regular, hyaline; hyphae very thin-walled, and without clamp-connections.

Distribution: reported from Ghana and Nigeria.

COPRINACEAE

COPRINUS (Pers. ex Fr.) S. F. Gray

This is a very well known genus whose species are characterised by the fact that the gills deliquesce on maturity, and the resulting drops of black fluid are often extruded on to the caps. The species grow on a wide variety of substrata such as dung, soil, sand, peat, on various fabrics, dead wood, charcoal, and on logs of palmwood. The cap when young is usually conical or campanulate, then expanding, margin or often the entire pileus deeply plicate-furrowed; stipe central, more or less straight; veil often forming an annulus or a volva, or absent; gills parallel, sinuate or adnexed, or adnate, mostly disappearing with age due to auto-digestion starting from the edge upwards; context usually white or whitish, fleshy or very thin or fragile to almost absent in very small specimens; hymeno-phoral trama regular; hyphal clamp-connections often present; hymenium consisting of isolated basidia arranged regularly among pseudoparaphyses; cystidia characteristically large, very frequent on the sides of the gills; cheilo-cystidia usually not differentiated; spherocysts large, loosely attached on the edges of the gills; basidia normal, rarely clavate, usually cylindrical, one to four-spored; spore-print black or fuscous; spores blackish or opaque, or transparent, but always deeply coloured, smooth, more rarely warty, echinate, reticulate, or angular, with a germ pore.

Distribution: cosmopolitan.

Practical importance: several species are considered as first class edible mush-rooms, but these should be harvested before the onset of autodigestion. When they are very young, some species are named 'weed fungi' because they develop as weeds in the manure beds prepared for edible mushroom growing. One species, *C. radians*, is reported to be the cause of fabric destruction, and another, *C. quadrifidus*, produces an antibiotic called quadrifidin.

COPRINUS DISSEMINATUS (Pers. ex Fr.) S. F. Gray

This is the most common species of the genus *Coprinus* and has been described many times. Some authors believe that this non-deliquescing fungus should have been placed in the genus *Psathyrella*. The fruit bodies are densely caespitose to gregarious, rarely solitary, and usually grow on and around the fallen stumps of trees and occasionally on the wood buried in the soil. Often the fruit bodies appear shortly after rain when thousands of these small mushrooms can be seen growing in a small area. The very young cap is ovate than becoming expanded campanulate, with a central disc, up to 10–15 mm in diameter; cap colour changing from whitish or pale buff to grey, except the disc which remains yellowish, fibrillose at first then becoming almost naked, splitting radially from above, flesh thick in the centre becoming very thin at the margin; gills thin, wedge-shaped, adnate, at first greyish then finally becoming black; stipe 2·5–4 cm long, whitish, thin, fragile, hollow, fibrillose, very slightly tapering upwards; spore-print black; spores pale brown, 7–9 (–11) × 3·5–5·5 μm, smooth, flattened, rather broad, with a distinct germ pore; basidia normal, four-spored, tetramorphic, the longer ones much protruding, narrow; pleurocystidia absent; cheilocystidia large, obtuse, scattered, developing from the edge and spreading up to about half way; pileocystidia very large, apex obtuse with an inflated base; spherocysts numerous; incrusted; all hyphae with clamp-connections.

Distribution: cosmopolitan.

COPRINUS SETULOSUS Berk. & Broome

The fragile carpophore of *C. setulosus* often grows on the soil among grass on lawns, and on other substrata such as vegetable refuse. Usually it appears during the early part of the rainy season. Cap conical bell-shaped, often becoming expanded up to 4–5 cm in diameter, with a central pale disc, ochraceous-buff, becoming paler towards the margin radially striate, surface containing short brown hairs, veil absent; gills extending as far as the stipe but not attached to it, narrow, first they are of the concolorous with the cap but become darker with age, non-deliquescent; stipe 2–11 cm in length, cylindrical and hollow, pure white and arising from a mycelial mat, annulus absent; context thin, hyaline; spores 8–10·5 × 7–8·5 × 6–7·5 μm, subglobose to ellipsoid, dark brown to black, thick-walled and with a small germ-pore; basidia clavate, four-spored; cheilocystidia absent; pleurocystidia numerous, hyaline and thin-walled; hymenophoral trama regular, hyaline, narrow, consisting of inflated thin-walled hyphae; all hyphae with clamp-connections.

Distribution: Ceylon, Ghana and Nigeria.

COPRINUS AFRICANUS Pegler (Fig. 32.)

This usually grows among fallen leaves and has a 3–6 cm high conical cap which finally expands to show a prominent umbo. The colour of the cap is grey

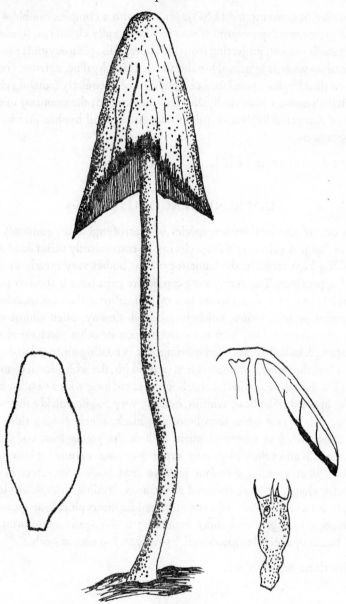

Fig. 32

at the centre but blackish at the margin as the gills deliquesce; veil absent but often a few silky fibrils are observed; gills free, becoming liquid after maturing; stipe 6–15 cm high, cylindrical, hollow, pure white, smooth; context thin, concolorous; spores 5·3–7·6 × 4–5 × 3·7–4·8 μm, ellipsoid, fuscus to black, discolouring in concentrated H_2SO_4, smooth with a complex double wall and a broad germ-pore; spore-print black; basidia broadly claviform, four-spored; cheilocystidia present, projecting from the young gills; pleurocystidia numerous, hyaline, thin-walled; hymenophoral trama regular, hyaline, narrow, consisting of thin-walled hyphae; pileal surface composed of irregularly radiating chains of elongated elements which are hyaline and thin-walled; the remnants of the veil consist of elongated hyaline or pale brown hyphae; all hyphae provided with clamp-connections.

Distribution: Uganda, Nigeria.

COPRINUS LAGOPUS (Fr.) Fries

This is one of the best known species of macrofungi and commonly occurs on horse dung. A culture of this species can be conveniently maintained on fresh horse dung kept moist in the laboratory; fruit bodies vary greatly in size and general appearance. The very young cap before expansion is about 0·5–20 mm high and 1–40 mm in diameter, at first cylindrical-oval then campanulate with a somewhat pointed umbo, radiately grooved, downy, often almost or quite naked, (the down having been removed by rain or other mechanical means), then flattened, radially sulcate and splitting, disc becoming more or less depressed, white at first then grey and finally black or blackish, the white down changing to small, thin, delicate, fugacious scales. Stipe 5–12 cm long with a swollen base and tapering upwards, fibrillose, whitish, hollow, very fragile, thickly tomentose at the base; gills at first white then becoming black, adnexed, very thin, parallel sided, deliquescing at maturity; pileus flesh in the young fruit body is white above the gills and yellowish-brown at the disc, when matured it becomes grey or black; pleurocystidia abundant in large fruit bodies, very few in smaller specimens, elongated oval, rounded at the apex, hyaline, completely bridging the space between the two adjacent gills; basidia dimorphic, four-spored; long basidia 40 × 8–10 μm, and short basidia 23 × 8–10 μm; spore-print black; spores black, opaque, elongated-oval, 15–16·5 × 8–9 μm, smooth.

Distribution: cosmopolitan.

PSATHYRELLA (Fr.) Quélet

Most of the species of this very widely distributed genus have a short life-span, the fruit bodies open rapidly, shed all their spores within a comparatively short period (from 12–20 hours) and then die. The species grow on soil, on dung, on dead or living wood, occasionally as parasites on living basidiomycetes, especially agarics. Cap often brown, campanulate, conical or vesiculose, usually distinctly hygrophanus, pellucid-striate when water-soaked, with a distinct cellular epicutis; gills narrow, wedge shaped, adnexed to adnate; stipe central, tubulose, flexuous or straight; with or without a veil, veil sometimes forming a ring but usually disappearing; volva absent; the base of the stipe sometimes connected to pseudorhiza or short rhizomorphs; context white or coloured; hyphae usually with clamp-connections; hymenophoral trama regular to sub-regular, pigmented or hyaline; cystidia often confined to the edges of the gills; basidia normal, occasionally two-spored; spore-print purplish-fuscous to deep-fuscous to black, more rarely a dull reddish; spores dark purplish-brown to almost black, of various sizes, small (below 6 μm) to large (up to 20 μm), smooth, with complex wall and germ pore.

Distribution: cosmopolitan.

Practical importance: edible; at least one species is reported to be an 'excellent' edible fungus.

PSATHYRELLA ATROUMBONATA Pegler

This small *Psathyrella* grows on forest litter. The cap is about 1·5–4·5 cm in diameter, conico-campanulate becoming expanded umbonate, light ochraceous-buff darkening at the centre and faintly striate at the margin; veil consists of loosely interwoven hyphae, hyaline, thin-walled, septate; gills attached broadly to the stipe and notched, crowded, pale grey, becoming dusky; stipe 5–9 cm in length, cylindrical, hollow, white and smooth; context very thin, white; spores 5·5–8·5 × 3·7 × 5·2 μm, dusky brown, translucent, germ-pore small; spore-print dusky; basidia claviform and four-spored; cheilocystidia abundant, hyaline, thin-walled, forming a sterile gill-edge, pleurocystidia absent; hymenophoral trama regular, hyaline in NH_4OH, consisting of thin-walled, inflated hyphae; all hyphae provided with clamp connections.

Distribution: reported from Uganda and Nigeria.

BOLBITIACEAE

BOLBITIUS Fries

Although the species of this genus are found almost throughout the world, *Bolbitius* is very little known. The fruit bodies usually grow on dung, soil, and on rotting trunks of wood. The cap is characteristically viscid, plicate-sulcate (as in some species of *Coprinus*) striate, fragile, the cuticle is cellular; stipe white or whitish, even near the base; gills splitting, after disintegration of the tissue following spore release; hymenophoral trama regular, cheilocystidia not abruptly capitate; basidia clavate, sunken and closely packed among the well-developed pseudoparaphyses, forming a paving on the hymenial surface, four-spored; spore-print dull brown; spores smooth, with germ pore, develop evenly over the surface of the gills.

Distribution: almost cosmopolitan.

Practical importance: none.

BOLBITIUS VITELLINUS (Pers. ex Fr.) Fries

This is a very delicate and variable species which usually appears singly or in clusters on dung or manured grass. Cap 1–7 cm in diameter, egg-yellow in colour, viscid, thin, membranous, striate at first, finally grooved, bell-shaped, soon expanded and usually liquefies after spore-discharge; gills pallid or brownish-ochre, crowded, thin, attached by a thread to apex of stipe, two sizes; stipe straw-yellow to whitish, very slender, fragile, more or less powdery floccose, becoming slightly wider towards the cap, readily separable from pileus, hollow, fragile, base white, myceloid; spore-print rusty-ochraceous; spores 12·5–14·5 × 7–9 μm, smooth, ellipsoid to sub-ovoid; basidia four-spored; paraphyses present, thin-walled; pleurocystidia absent; cheilocystidia abundant, thin-walled; hymenophoral trama of enlarged cells more or less parallel; context consists of enlarged hyaline and interwoven hyphae.

Distribution: cosmopolitan.

Practical importance: reported edible, but not worth trying.

AGROCYBE Fayod

Some species of the genus *Agrocybe* are very common in the tropics, usually growing on such substrata as dung, seeds, decaying wood and manured soil. The species appear during the rainy season in gardens, lawns, fields and meadows. Cap fleshy, entirely or partly striate, with an epicutis consisting of spherical or small pear-shaped cells; stipe white or coloured, smooth or rough, often with a well-developed persistent ring, base of the stipe often attached to thin white rhizomorphs; gills broad; cystidia present or absent; cheilocystidia always present; basidia two-, three-, or four-spored, otherwise normal; spore-print dark coloured; spores smooth, chestnut coloured, with thick double walls; germ pore broad and truncate, or narrow and non-truncate, often indistinct, occasional spores have two or three germ pores; hyphae with or without clamp-connections.

Distribution: almost cosmopolitan.

Practical importance: several species are excellent edible mushrooms; some are cultivated, though by a rather primitive method. *A. dura* contains an anti-biotic, agrocybin.

AGROCYBE BROADWAYI (Murr.) Dennis (Fig. 33.)

A common mushroom in grassland areas, it usually appears during the rainy

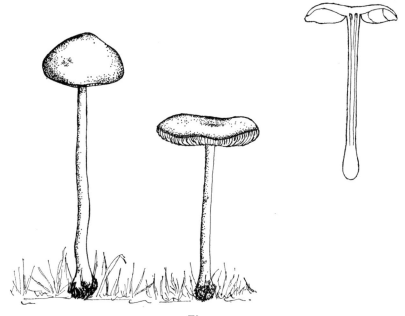

Fig. 33

season. Cap is 0·5–3·5 cm in diameter, white or alutaceous cream, subviscid, but cracking near the edges; gills are either attached broadly to the stem or extend as far as the stem but are not attached to it; stipe 2·5 to 5 cm in length and 2–4 mm in diameter, hollow; veil absent; context thin, white, consisting of densely interwoven, very thin-walled hyphae; spores ovoid to broadly ellipsoid, 11·0–15 × 6·5–9 μm, rusty-brown under the microscope, smooth, thick-walled with a germ-pore; basidia clavate, two-, three-, or four-spored; cheilocystidia present, thin-walled, hyaline, varying in shape from subglobose to clavate, and form a heteromorphous gill-edge; pleurocystidia absent; hymenophoral trama regular, hyaline, consisting of thin-walled hyphae; sub-hymenial layer well-developed and subcellular; surface of the cap is an epithelium of pedicellate spherocysts which are subglobose, piriform or clavate, hyaline and thin-walled or with a slightly thickened, brown pigmented wall; pilocystidia absent; all hyphae have clamp-connections.

Distribution: this species was previously known in tropical America, but later it was also recorded in Trinidad, Ghana and western Nigeria.

CORTINARIACEAE

GYMNOPILUS Karsten

Species of the genus *Gymnopilus* grow on living or dead trees, living orchids and the roots of the grasses. Cap bright coloured, usually yellow or red, pigment incrusting the hyphal walls, glabrous fibrillose, squamulose, squarrose, floccose or rimose; epicutis formed by hyphal chains which are frequently erect, forming a kind of trichodermium, at least in the centre of the cap; gills adnexed to decurrent, narrow to broad, becoming very brightly and richly rust-coloured in mature carpophores; spore-print amber-brown; spores coloured, double-walled, ellipsoid, warty, rough when seen under an oil-immersion lens; basidia

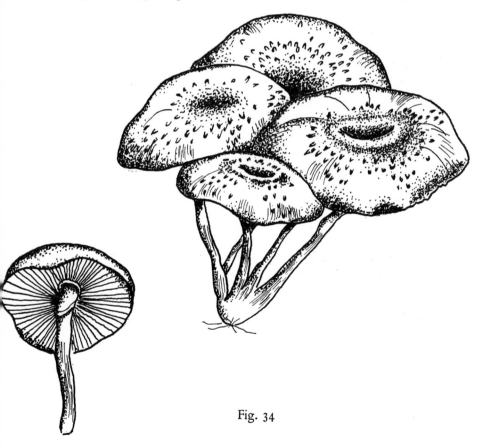

Fig. 34

usually clavate, four- or two-spored; marginal cystidia hair-like, facial cystidia usually absent; hymenophoral trama regular; stipe usually yellow, equal or longer than the diameter of the cap; veil present and in some species remains attached to the stipe as a well-developed annulus; context often bitter; hyphae with clamp-connections; the pileus becomes black with KOH, especially when it is covered with spores.

Distribution: almost cosmopolitan; rare in temperate and arctic zones.

Practical importance: none of the species has much economic importance at present, although a few species may cause mild destruction of wood. *G. aculeatus* apparently produces the endotrophic mycorrhiza of a tropical orchid (Fig. 34.). The yellow soluble pigment of many species is used as a dye in cytological work. It is suspected that a few species are poisonous.

GYMNOPILUS PAMPEANUS (Speg.) Singer

This brightly coloured mushroom is a common tropical species and usually grows in tufts on the stumps of trees (particularly in Eucalyptus plantations), during the rainy season. Cap 2–12 cm in diameter, at first convex then expanding, depressed at the centre, surface dry, ochraceous - buff becoming pale orange-yellow at the margin, ornamented with small, appressed fibrillose and minute scales which are usually crowded into the caps centre, margin entire; gills almost decurrent, moderately crowded, with lamellulae of two different lengths, edge entire; stipe 3–12 cm long, cylindrical, concolorous with the pileus, tapering upwards, tough, solid, surface striate, fibrillose; annulus membranous, fleshy, persistent; context firm-fleshy, consisting of interwoven, thin-walled, hyaline hyphae; spore-print ferruginous; spores 7–10 $\times$ 5–7 μm, ellipsoid, inamyloid, thick-walled, germ pore absent; basidia clavate, four-spored; cheilocystidia hyaline, thin-walled, fusiform; pleurocystidia absent; hymenophoral trama regular, hyaline, inamyloid, consisting of parallel thin-walled hyphae; subhyemnial layer well developed, consisting of interwoven hyphae; all hyphae with clamp-connections.

Distribution: mostly tropics; common in Uganda, Kenya, Tanzania and Nigeria.

BOLETACEAE

PHAEOGYROPORUS Singer

Species belonging to this genus are frequently grow on the ground, often form-ing mycelial crusts around the roots of trees. Most of the species are tropical. Pileus pulvinate, planoconvex and centrally depressed, fleshy, brownish-olive, hymenophore tubular; spores of the tube minute to medium sized; hymeno-phoral trama bilateral; spore-print light brownish-olive; spores short and cylin-drical; all hyphae with clamp-connections.

Distribution: South America, tropical Africa, Ceylon, Brazil and Argentina.

Practical importance: some species form destructive mycelial crusts around the roots of citrus trees.

PHAEOGYROPORUS PORTENTOSUS (Berk & Br.) McNabb

This species is widely distributed in the tropics and pops up on lawns and around forest trees during the rainy season. The cap size varies from 6–50 cm in diameter, adnexed to depressed in the centre, at maturity it often looks like a horse's saddle, burnt amber in the centre gradually becoming lighter towards the margin, cuticle non-separable, surface smooth or cracked; hymenophore tubular, adnexed to free, tubes longer in the middle becoming shorter towards the margin and the stipe, yellow turning brownish when cut; pores irregular, angular; stipe eccentric, up to 20 cm in length and about 5 cm thick, bulbous at the base; context thick, yellow, soft when fresh; hyphae hyaline, inamyloid, interwoven; spores 6–9 × 4–7 μm, ellipsoid, inamyloid, thick-walled; basidia clavate, four-spored; hyphae clamped; sulphuric acid turns the surface of the pileus and the stipe scarlet.

Distribution: Ceylon, Sierra Leone, Kenya and Nigeria.

Practical importance: it is believed by the local people that it causes intoxi-cation when eaten.

PHLEBOPUS (Heim) Singer

The species included in the genus *Phlebopus* are exclusively tropical and usually grow on the soil in marshy places. Fruit bodies are usually very large, and quickly putrefy. Cap convex, thick, margin often becoming very narrow; gills absent; hymenium poroid, tubes not firmly fixed to the pileus, separating easily; stipe eccentric, very tough, frequently bulbous, furrowed at the base, and attached to a pseudosclerotium; context soft, watery, generally yellowish and more or less blue.

Distribution: tropics.

PHLEBOPUS SILVATICUS Heinemann (Fig. 35.)

This is a common species of the tropics which grows on the soil, usually on the slopes in dry forest. Cap thick up to 30 cm diameter, convex at first then becom-

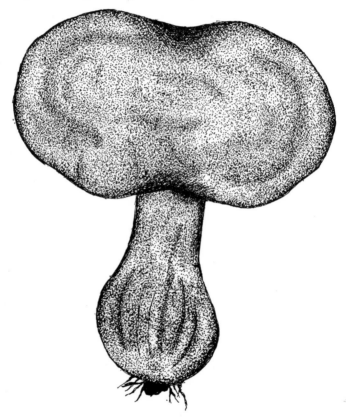

Fig. 35

ing plane, with margin turned up. The cuticle of the pileus is thick, orange coloured, turning violet at old age, membranous, peeling off as a crust; scales large and thick, light brown to ochraceous; stipe eccentric, clavate, bulbous, up to

9 cm long and 3 cm wide, claviform and distinctly furrowed at the lower part, covered with a cuticle of the same colour as that of the cap, cuticle detaching itself as fruit body becomes old; the stipe extends into the soil to a pseudosclerotium; hymenophoral tubes can be separated from the pileus flesh, concave, becoming partly joined, deep yellow, occasionally green, shallow, pores fairly large, concolorous; context viscid, soft ochraceous, becoming brown at the base of the stipe and turning dark blue when cut; spores thin, ellipsoid, subglobose, 6·5–8·5 × 5·5–7·5 μm, non-amyloid.

Distribution: tropical.

PODAXACEAE

PODAXIS Desvaux

This peculiar tropical fungus is fairly common but the taxonomy of the genus seems to be in a confused state. So far, almost all authors seem to agree with Morse (1933) who has suggested that all specimens of *Podaxis* belong to a single species, but it seems unlikely that this is the true position. Ingold (1965) has suggested that in Ghana at least two clearly defined species are encountered.

The fruit body of *Podaxis* has an ellipsoid head, supported on a slender stipe, and resembles certain mushrooms, however, the texture of the sporocarp is entirely different. The cap is supported on a firm, rigid and almost woody stipe; exoperidium covers the whole sporocarp, breaking into scales; endoperidium tough, dehiscing by splitting, sometimes bending upward slightly to expose the base of the glebal mass; gleba copious, developing in between the upper part of the stipe (columella) and the endoperidium.

Distribution: cosmopolitan; widespread in tropical and subtropical zones.

PODAXIS PISTILLARIS (L. ex Pers.) Fr. Sensu Morse (Fig. 36.)

The description of *P. pistillaris* given here is based on the specimens collected in Nigeria and observed growing on the surface of small termite-hills in Shagunu near the Kainji lake area.

The overall length of the sporocarp may range up to 25 cm; cap ellipsoid to subcylindrical or subconic, 10–13 cm long, and 3·5 cm in diameter. Exoperidium thin, pale, fawn, breaking up into scales, more or less deciduous; exoperidium dirty-white to fawn, thin, woody, dehiscing by breaking away from the point of attachment to the stipe, and then splitting into a small number of rays which bend outward and then upward, and finally detaching completely; stipe woody, tapering upwards, projecting into the cap and forming a pseudocolumella, bulbous at the base, hollow, scaly, often rooting by a conspicuous hyphal strand; gleba at maturity becomes a mass of reddish-brown powdery spores, which develop among the coiled, thread-like structures (capillitium); (these threads arise from the stipe and pass outward to meet the peridium without actually being attached to it); capillitium strands often 2–3 cm long, completely unbranched, nonseptate, later spirally thickened; spores ellipsoid to sub-spherical,

7–20 µm × 5–15 µm, reddish-brown, thick-walled with gelatinous envelope and a well-marked germpore.

Distribution: widespread in tropics and subtropics.

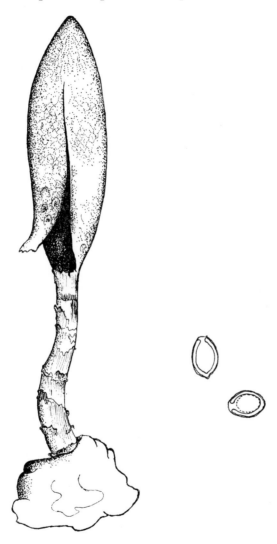

Fig. 36

PHALLACEAE

MUTINUS Fries

The fruit bodies of this genus are usually brilliantly coloured; stipe rosy-pink, hollow, cylindrical or fusiform, smooth or regulose; receptacle not free, but an extension of the stalk, deep red in colour, smooth, covered with glebal mass; gleba mucilaginous, olivaceous with a foetid smell; spores elliptical, smooth, chlorohyaline.

Distribution: cosmopolitan.

MUTINUS BAMBUSINUS (Zollinger) E. Fischer

Mutinus bambusinus is a rather variable species, the following description is based on a specimen collected from the Biological Garden, University of Ife, Ibadan.

Egg white in colour, ovoid, strongly rooted by a basal strand of hyphae, dehiscing by an apical slit; receptacle 12 cm high, hollow, cylindrical, white below shading to pink at the top; wall of the receptacle one chamber thick, the chambers opening internally; head of the receptacle fertile, crimson, hollow, conical, chambered in the same manner as the sterile part of the receptacle but more robust, faintly regulose on the surface; gleba covering the head, olivaceous, mucilagenous, strongly smelling but not unpleasant; spores elliptical, smooth, chlorohyaline, 5 μm × 2 μm in size.

Distribution: Africa, Asia, East and West Indies and South America.

PHALLUS L. ex. Persoon

The fruit body of this genus consists of a hollow stipe with a cup-shaped volva at the base and a glebiferous cap at the apex. The wall of the stipe is spongy with a single or double layer of chambers; the chambers open either to the interior of the hollow stipe or to the exterior. The colour of the stipe is white, reddish or orange. The cap is white or occasionally orange, perforated at the apex, and consists of a thin basal membrane thickened in some parts. The thickened parts may be regular, papillate, tuberculate or reticulate; gleba borne on the surface of the cap, olivaceous, foetid, mucilaginous.

The genus contains some species in which the indusium hangs down from near the apex of the stipe and forms a network when fully expanded. It has been a practice to segregate such species into a separate genus *Dictyophora*. However, it has been suggested (Dring, 1964) that the indusium alone is an inadequate criterion on which to base a genus.

Distribution: widespread; throughout the tropics and subtropics.

PHALLUS INDUSIATUS Vent. ex Persoon (Fig. 37.)

This is a beautiful white and common tropical species which can be easily identified by its characteristic foetid smell and its well-developed indusium which is

Fig. 37

attached to the upper end of the stipe just under the point of attachment of the fertile cap, and hangs down around the stalk like a skirt of white delicate network. The immature fruiting body with a reddish-brown shade is an egg-shaped structure. At maturity the outer layer of the egg splits open at its apical region and a hollow, spongy stalk grows out to about 15 cm high. The outer covering of the egg remains attached to the bottom of the stalk and forms the volva. The bell-shaped, conical pileus, with a perforated apex, is attached to the top of the stipe and is thrown into strong folds which form a network on the surface. Gleba is formed on the folded surface of the pileus and is olivaceous, foetid and mucid; spores ovoid, smooth, chlorohyaline, 3·5 μm × 1·5 μm.

Distribution: though this species is commonly found in the tropics, other vividly coloured forms have been reported from other countries and these have other names. The specimen described here (I.M.H. No. 413) was collected from a forest at Idanre Hill, Carter Peak, at an altitude of about 1700 feet, on 20th April, 1969.

PHALLUS RUBICUNDUS (Bose) Fries

This is a common tropical fungus which grows on the ground in forests or in the open. The species is very variable both in colour and habit, and is reported from many parts of the world. It has been described by at least 15 different specific names. The following description is based on the material collected in June 1960 from the University of Ife Biological Garden, Ibadan.

Egg globose, rooted by a single hyphal strand; peridium whitish, dehiscing by apical splitting; stipe red, 16 cm long, spindle shaped, hollow, spongy; cap red, bell-shaped, easily detachable from the stipe, imperforate, surface of the cap is finely folded; gleba olivaceous, mucilaginous, strong foetid smell, enabling the specimen to be detected from a fair distance; spores ovoid, cylindrical, 5 μm × 2·5 μm.

Distribution: throughout the tropics and subtropics.

CLATHRACEAE

CLATHRUS Mich. ex Persoon

The genus *Clathrus* includes some of the most beautiful fungi, unfortunately the odour of the fruit body is so strong, unpleasant and nauseating that most students of mycology prefer to admire them as preserved specimens in tightly sealed collecting jars. These fungi frequently grow in garden borders and on heaps of leaves, and usually appear when rains follow a warm dry period. The receptacle consists of a more or less spherical network, hollow, sessile and whitish, orange or red in colour. The spore slime may cover the entire inner surface of the receptacle or be restricted to the intersections of the arms of the network. The arms may be tubular, or triangular and show a more or less regular arrangement of chambers, though this arrangement may not be noticed in large and spongy species.

Distribution: tropical and temperate zones, commonly in Europe, Asia, Africa, West Indies and Japan.

Practical importance: none, though sometimes in France the peasants regard *Clathrus* as a cause of cancer and it is often thought to produce skin eruptions if handled.

CLATHRUS PREUSSII Hennings (Fig. 38.)

This species is quite common in the tropics, but it is extremely fragile and therefore not easily collected as an entire specimen. The young fruit body is like a soft egg and usually grows just below the soil or under leaf litter. The base of the egg is attached to a rhizomorph which can be traced down to buried wood. The outer covering of the egg (peridium) splits irregularly and soon after the receptacle emerges to quickly grow upward. The base of the receptacle remains enclosed in the volva. The mature receptacle consists of a more or less spherical white network, about 6 cm in diameter, the meshes of roughly equal area are formed by triangular and chambered arms. In cross-section the arms exhibit a definite pattern of chambers: a large chamber on the inner side, three medium-sized chambers in the middle and the two smallest, often fringed, on the outer side with a slight groove between them. The mucilaginous, greenish glebal mass spreads over the inner surface of the arms, and emits a strong and unpleasant odour; spores 2–5 μm $\times$ 1·5–2 μm, ellipsoid, faintly greenish and smooth-walled.

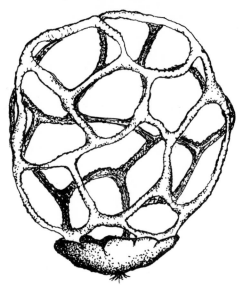

Fig. 38

Commonly the species is known as the 'white clathrus of Africa'. The white colour, characteristically chambered arms, and the typical shape of the meshes serve to distinguish this from all other species.

Distribution: tropics, mostly in Australasia and Africa.

KALCHBRENNERA Berkeley

The species belonging to the genus *Kalchbrennera* are not very common; occasionally they are found growing on well-manured soil and can be identified by the deep colour of the receptacle and the ornamented cap. Eggs more or less spherical, dehiscing by apical splitting; receptacle emerging from the torn peridium, consisting of a hollow stipe with a hemispherical network at the top; simple or forked processes are projected out from the network and spread all around; gleba mucilaginous, foetid, olivaceous, spread on the outer surface of the network, between the processes. The genus contains only a single species.

Distribution: tropical; mostly in Africa.

KALCHBRENNERA CORALLOCEPHALA (Welw. & Curr.)
Kalchbrenner (Fig. 39.)

This fungus is usually encountered growing on the ground, in woodland or in the open. Although it is variable with regard to the length of its appendages, it is nevertheless easily recognised. The egg is more or less spherical and grows up to

4 cm in diameter; receptacle stipitate; network of the receptacle bright red, hemispherical, up to 3 cm in diameter, meshes varying in number; arms transversely rugulose, tubular, 3 mm in diameter; appendages bright red, branched or simple, projecting 2 cm high from the network, often smaller; stipe cream coloured, often pinkish at the top, cylindrical, 12 cm long and about 3 cm in diameter, hollow with chambered walls; gleba olivaceous, mucid, foetid; spores ellipsoid, smooth, chlorohyaline, 3·5–4·5 μm × 1–2 μm.

Distribution: Africa; common in Nigeria, Ghana and Kenya.

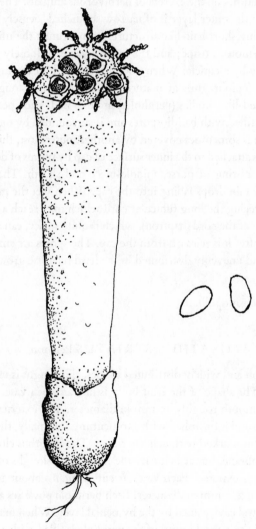

Fig. 39

NIDULARIACEAE

CYATHUS Haller ex Persoon

The vase-shaped fruit bodies of the genus *Cyathus* are commonly known as 'birds nest' fungi and can be found growing on dead wood, or soil in contact with wood, and sometimes on the excreta of herbivorous animals. The wall of the cap is three-layered: the outer layer is of narrow, branched, densely woven hyphae sometimes bearing short hair-like structures (tomentum); the middle layer is of pseudoparenchymotous tissue; and the inner layer is of loosely woven hyphae bound internally by a cuticle. When young the mouth of the cap is closed by a thin membrane (epiphragm), at maturity this membrane sloughs off and the lenticular, or seed-like, bodies (peridioles) are exposed. Each peridiole contains a central cavity filled with basidiospores and is surrounded by one or two layers of cortex which is sometimes covered by another, outermost, thin hyaline layer (tunica) which is attached to the inner surface of cap by means of delicate mycelial connections, consisting of purse, middle-piece and sheath. The peridioles are dispersed by the rain drops falling into the cap. When wet the purse ruptures at its lower end, freeing the long funicular cord which may reach a length of up to 20 cm. The base of the cord (haptron), which is quite sticky, can adhere easily to nearby objects after it is released from the cap. The spores are smooth, ellipsoid, thick-walled, and unevenly distributed in the hard sclerotic ground tissue of the peridiole.

Distribution: cosmopolitan.

CYATHUS STRIATUS Persoon

A fairly common and widely distributed species, this fungus is usually found on rotting wood. The shape of the fruit body is like an open vase, 5–15 mm high with a circular mouth roughly 10 mm in diameter. The external surface of the vase is brownish and is furnished with a tomentum. Internally, the vase is a shiny grey colour and is marked with delicate ridges. The mouth is closed at first by a thin white membrane, but at maturity the membrane sloughs off and the inner part of the vase is exposed. Each vase, or cup, contains about 10–12 grey eggs (peridioles) about 2–3 mm in diameter. Each peridiole possesses a wall of several layers and a central cavity lined by the hymenial layer. When mature the basidia disintegrate and the central cavity of the peridiole is filled with a mass of spores.

Each peridiole is attached to the inner surface of the vase by a short stalk, 2–3 mm long, which is differentiated into an upper part (purse), a middle-piece, and a lower part (sheath). The purse is delicate, hollow, tubular, closed at the lower end, and coiled within, containing a 5–10 cm long 'rope' of hyphae (funiculus); the upper end of the funiculus is firmly attached to the peridiole and the lower end is attached to a mass of very sticky hyphae (haptron).

Distribution: cosmopolitan.

CYATHUS POEPPIGII Tulasne

The fungus is usually encountered growing on wood or on soil in contact with wood. Cups dark brown, obconical, 8 mm long and 8 mm in diameter, furrowed on both inner and outer surfaces. The furrows are sometimes obscured on the external surface of the hairs. Peridioles black, shiny, up to 2 mm in diameter, tunica absent, cortex two layered, dull brown; spores ellipsoid, 10–15 × 17–20 μm or larger.

Distribution: very common in tropics and subtropics.

CYATHUS LIMBATUS Tulasne

The fruit bodies are usually encountered growing on wood or on soil in which wood is buried. The cup is dark brown, attains a length of up to 10 mm and a diameter of up to 8 mm; both outer and inner surfaces of the cup are ornamented with vertical grooves, hair-like structures occur on the outer surface of the cup which at an early stage of development cover the grooves. The base of the cup is attached to a tuft of mycelia which forms a pad-like structure on the substratum. Peridioles are lentil-shaped, deep brown, shiny, about 2 mm in diameter with a two layered cortex; the outer layer is very thin, often consisting of a single layer of red-brown hyphae. Spores usually 10–15 × 15–20 μm, ellipsoid, thick-walled, smooth and hyaline.

Distribution: throughout the tropics and subtropics.

SPHAEROBOLACEAE

SPHAEROBOLUS Tode ex Persoon

This genus includes an uncertain number of species and is of worldwide distribution. Its very delicate and wonderfully constructed fruit bodies are commonly found growing in groups on rotting wood, on dung, on old saw-dust and on rotten twigs and leaves. The immature fruit bodies are pale yellowish or whitish, spherical, and are immersed in white cottony mycelium. At maturity, the outer wall splits from the top into four to eight sharp pointed teeth, which bend outwards exposing the single, spherical, orange-yellow, spore-containing ball. Six layers can be distinguished in the young peridium but at maturity only two remain distinct, separated by an air space and joined at the extremities of the points – as if it were one cup within another, the inner cup containing the glebal ball (peridiole), immersed in a slimy liquid. The glebal ball is about 1·4 mm in diameter, covered by a thin, adhesive brown skin, and consists of tens of thousands of thick-walled spores and thousands of gemmae embedded in the tough matrix. The spores are at first borne on basidia but at maturity the basidia disintegrate and the centre is filled with a mass of spores. The gemmae are larger, oval or oval-elongate, pieces of vegetative hyphae and gemmae are capable of germination. Suddenly, due to tissue tension, the inner cup turns inside-out and the peridiole is catapulted with such force that it may travel vertically a distance of several metres. The outer, non-everting, membrane holds fast the inner membrane by its teeth and prevents it from ejecting after the peridiole.

Several species have been proposed, but these seem to differ only in habitat; all are now treated as varieties of one species.

Distribution: almost cosmopolitan.

SPHAEROBOLUS STELLATUS Tode ex Persoon (Fig. 40.)

This is one of the smallest species among the Gasteromycetes, and usually grow on wood or occasionally on soil or dung. The fruit body is not more than 2 cm in diameter, gregarious, pale ochraceous, usually growing from a mycelial sheet when on wood, or strongly rooting from the base when on soil. The peridiole (spore-containing ball) is surrounded by a peridium of six distinct layers. When mature, the peridium splits along a number of lines radiating from the apex thus carving out four to eight recurving rays which are often bright orange-yellow

Fig. 40

on the adaxial side. At the same time the peridium separates into two stellate cups, one fitting inside the other, and remains in contact only at the tips of the rays. The outer cup is composed of the three outermost layers of the peridium and the inner cup of the two middle layers, the inner layer consisting of a palisade tissue of relatively large cells. The outer layer is of fine, interwoven, tangential hyphae, the sixth innermost layer autolyses to produce a lubricating fluid. Peridiole chestnut-coloured, 1·4 mm in diameter, loosely fitted in the inner cup, and is immersed in the lubricating fluid; spores thick-walled, smooth, hyaline, elliptical, 6–10 × 4–5 μm, and are mixed with, swollen, densely protoplasmic, dikaryotic, thread-like tubes which are said to be capable of germinating.

Distribution: cosmopolitan.

LYCOPERDACEAE

LYCOPERDON Persoon

Commonly known as 'puff-balls' this genus is so named because of the method of spore dispersal and ball-shaped appearance. The size ranges from about 1·3 cm to 31 cm or more across. They usually grow on tree stumps, on decaying logs, or on the ground in the woods, and can also be seen growing on lawns, golf-greens, and grassy open fields. It has been reported that all species of puff-balls are edible. However, they should be collected for cooking when they are young and are pure and white inside. As the fruiting body matures, the centre begins to turn yellow, by which time the flavour is spoiled.

The fruit bodies possess a central fertile region covered with two distinct layers; the outer layer (exoperidium) is of simple or compound spines or of branny or scurfy particles, occasionally velvety, it may remain intact or cracks in the upper portion and falls away; the inner layer (endoperidium) is thin, tough, persistent, dehisces by an apical pore, the dry spores are 'puffed' out when an object strikes the membrane. At maturity the central fertile region is converted into gleba, a powdery mass of spores intermingled with long thread-like hyphae known as capillitium which is derived from the trama, often originating in a central tuft or pseudocolumella, attached distally to the inner surface of the endoperidium, dry, springy, simple or branched, sometimes undeveloped; spores generally coloured or ornamented with spines or warts.

Distribution: cosmopolitan.

Practical importance: the fruit bodies of most of the species are edible when young. The mass of spores and capillitium were at one time used for staunching blood flowing from wounds.

LYCOPERDON PUSILUM Batsch ex Schumacher

Commonly grows on the ground in open places. Fruiting body globose, about 2 cm in diameter, sterile base and diaphragm is absent, it attaches itself to the substratum by means of a strong tuft of mycelia. Exoperidium fragile, fragmenting into superficial warts; endoperidium membranous, smooth, dehiscing by an irregular pore; gleba yellowish becoming brown with age; capillitium freely branched and sometimes septate, fragmenting, irregularly shaped; spores globose, 3–5 μ in diameter, having minute wart-like protuberances and the remains of a pedicel.

Distribution: cosmopolitan.

LYCOPERDON PRATENSE Persoon

This is a common puff-ball of the tropical countries and appears frequently on house lawns during the rainy season. The fruit body grows up to 5 cm in diameter, subpulvinate to obconical and folded into pleats at the base. The exoperidium is fragile and fragmented into faintly coloured, compound spines; endoperidium buff or grey, at maturity it dehisces by a definite apical stoma which sometimes becomes torn with age and weathering. The species is characterised by a sterile and chambered base which is separated from the gleba by a conspicuous diaphragm confluent with the endoperidium. Gleba yellowish turning to brownish black. Capillitium almost absent except at periphery of gleba where it is represented by a few straight, occasionally branched and septate hyphae. Spores 3·5–5 μm, in diameter, smooth to finely spiny.

The species is recognised by its diaphragm and capillitium and is sometimes placed in a separate genus *Vascellum* and often called *V. pratense* (Pers.) Kreisel.

Distribution: Africa, Australasia and Europe.

CALVATIA Fries

The genus includes most of the larger puff-balls and its species commonly grow in grassy places and in forests, mostly appearing during the rainy season. The fruiting bodies are medium to large, depressed globose to pyriform, with a sterile base which is attached to the substratum by a strong mycelial strand. Both the peridial layers are quite thin and fragile. When mature both layers are fragmented in irregular patches at the apical region, and the powdery gleba is exposed. Sterile base chambered or compact, diaphragm when present tending to develop late. Gleba powdery, copious; capillitium threads are long, and sometimes break into smaller pieces, sparingly branched, occasionally septate; subgleba ranging from small and dense to massive with very large chambers; spores globose to broadly ellipsoid, smooth to echinulate.

Distribution: mostly in temperate and tropical zones.

Practical importance: recent work with *Calvatia* indicates that it may contain an anticancer substance (calvacin), thus confirming the beliefs of North American Indian tribes that eating these fungi prevents stomach tumours.

CALVATIA SUBTOMENTOSA Dissing & Lange (Fig. 41.)

This species usually grows during the rainy season on the ground in forest zones. The fruit body is pear-shaped, the basal part of the fruit body is sterile and occupies about one-third of the entire volume. Exoperidium thin, velvety, dark

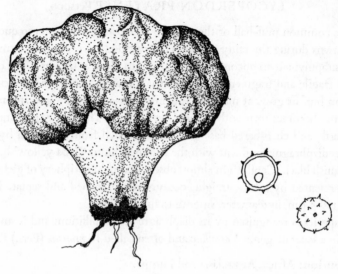

Fig. 41

brown; endoperidium thin, light yellow; both peridial layers are fragile and split into large fragments which eventually fall free from the apical part of the fruit body; gleba greenish-yellow; spores globose, 3·5–5 μm in diameter, and ornamented with sharply pointed spines; capillitium much branched, septate, fragmenting, subglebal region minutely chambered and lighter in colour than the gleba.

Distribution: mostly tropics, widespread in Congo, Kenya and Nigeria.

CALVATIA LONGICAUDA (Henn.) Lloyd
(= C. AGARICOIDES Dissing & Lange)

This species is commonly found growing in shady forests and is recognisable by its mushroom-like shape. Some of the spinning top-shaped fruiting bodies are sometimes mistaken for *C. gardneri* but they differ in possessing obviously chambered subgleba and irregularly branched capillitial hyphae which lack the characteristic holes of *C. gardneri*. The specimens possess a round, flat or concave-topped head which is abruptly differentiated from the stipe and reaches about 8 cm in diameter. Stipe subcylindrical, slightly tapering towards the base, 4–5 cm long; exoperidium slightly velvety; endoperidium very thin, fragmenting and falling away when drying; gleba rusty-brown; capillitium pale-yellow, much branched, often swollen at the septa, pits absent; spores 3–4 × 3·5–4·5 μm, sub-globose to ovoid or ellipsoid, minutely spiny, pale yellow with a short pedicel; subgleba brown, chambers 0·5–1·0 mm in diameter.

Distribution: Cameroon, Congo, Uganda, Madagascar and Nigeria.

CALVATIA GARDNERI (Berk.) Lloyd
(= LYCOPERDON GAUTIEROIDES Berk. & Br.)

This is a common *Calvatia* of tropical west Africa. The fruit bodies are turbinate ranging to about 10 × 10 cm in size. A fairly large portion, about half the length, is occupied by the sterile base. Exoperidium is generally chestnut coloured, breaking into small fragments; endoperidium ochraceous becoming rusty above, paler below, fragile; gleba ochraceous, fragile; capillitium branched, septate and honey coloured; spores globose 3–4·5 μm in diameter, rough, possessing delicate spines; subgleba at first ochraceous becoming brown, chambered; diaphragm absent.

Distribution: tropics; mostly in Africa, East Indies, Mauritius and Ceylon.

CALVATIA CYATHIFORMIS (Bose) Morgan (Fig. 42.)

This is a common species of west Africa and grows frequently in groups among grass in open places during the rainy season. It can be easily recognised by its large size. Just before dehiscence, the external surface of the fruit body becomes wet and both exoperidium and endoperidium become aeriolate and finally break into small fragments. The base remains attached to the ground as a dark purple-brown, spongy, conical, collared structure which if it is not disturbed by external agents, remains intact for many months after the gleba and peridium have dispersed.

The fruiting body is almost spherical to pear-shaped, up to 15 cm in diameter and possesses a well developed sterile base. Exoperidium thin, smooth or cottony, often breaks into fragments, cream at first becoming brown or greyish yellow, flesh becomes black on bruising; endoperidium brown at maturity, thin, fragile and breaks with the exoperidium; gleba greyish-purple; capillitium threads

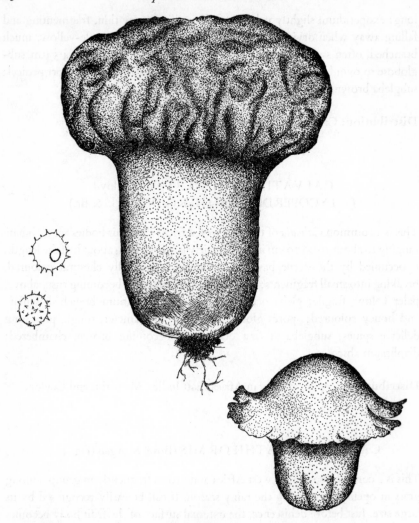

Fig. 42

branched, septate, fragmenting with minute pits, pale olivaceous; spores spherical, shortly pedicellate, furnished with hyaline echinules, 4–6 μm in diameter.

Distribution: widespread in the tropics and subtropics.

CALVATIA FRAGILIS (Vitt.) Morgan

This *Calvatia* differs little from *C. cyathiformis* and, according to Dring (1964), it is convenient to group these together with *C. lilacina* as a single species, demoting the various members to subspecies. Purely for the sake of clarity of presentation it is described here as a separate species. It differs from other subspecies in having medium sized fruiting bodies up to 8 cm in diameter, obovate to pyriform. Sterile base compact to minutely and obscurely chambered, occasionally with a few large chambers near the base. Spores spherical, 4–5 μm in diameter, long spined, purple-brown in colour.

Distribution: throughout the tropics and subtropics; North America to Ontario, central and southern Europe.

GEASTRUM Micheli ex Persoon

Species of *Geastrum*, commonly known as 'earth-stars', are frequently found hidden under leaf litter. The immature fruit body is globose to accuminate, hypogeal or epigeal. At maturity it breaks radially along several lines of weakness from the apical region, and the endoperidium is exposed. When fully open, the entire fruit body presents the characteristic 'star-shaped' appearance. Exoperidium consists of three layers; the outer mycelial, the middle fibrous, and the inner fleshy. Endoperidium depressed globose to ovoid, pedicellate or sessile, of a single thin, papery layer, smooth or slightly tomentose, dehiscing by an apical pore through which the spores are released; peristome present or absent, the properties of the peristome are important in distinguishing the species; gleba copious, powdery; capillitium of long, simple pointed hyphae; spores globose or subglobose, coloured.

Twenty species have been reported from the Congo and are described by Dissing & Lange (1962). The following are the more common species and can be identified easily.

Distribution: cosmopolitan.

GEASTRUM VELUTINUM Morgan (Fig. 43.)

This species is easy to distinguish on account of its large size and its harshly felted, paler, thicker, more pliable and continuous mycelial layer, and smaller, more finely verrucose, darker spores. The fruit bodies commonly grow on the ground, or on wood and other plant debris in forests. The immature, unopened sporocarps are either spherical or egg-shaped, occasionally slightly umbonate at the top, attached to the substratum through a basal strand of hyphae. Exoperidium splits into 5–8 expanded broad, thick, nearly unequal rays; the inner layer of the exoperidium is fleshy, moderately thick, adnate, flesh coloured, becoming amber to dark brown when dry and usually cracking across the base of the rays to reveal the middle fibrous layer; the outer mycelial layer, thick, pliable, very finely felted, buff ochraceous to amber, coarse to the touch, free from debris, adnate, occasionally separating and bending away from the fibrous layer giving the appearance of a double whorl of rays. Endoperidium smooth, sessile, more or less spherical, up to 2 cm in diameter, the lower part is enclosed by the sac-like base of the exoperidium; peristome apical, broadly conical, fibrillose, depressed

Fig. 43

round the edge, concolorous, or paler; gleba amber; pseudocolumella cylindrical to spindle-shaped; capillitium threads occasionally branched; spores globose, 2·5–4·2 μm, fuscous, minutely to finely, occasionally moderately, verrucose.

Distribution: widespread; most common in Africa, America, Asia, Australasia and Pacific Islands.

GEASTRUM TRIPLEX Junghuhn (Fig. 44.)

The fruit body is usually found growing on humus-rich soil or on the debris of plants under shady places in forests, and can be easily distinguished by the characteristic cracking of the fleshy layer around the base of the rays to leave a 'cup' enclosing the base of the endoperidium. The immature sporocarp grows up to

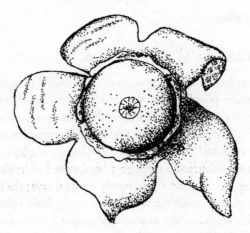

Fig. 44

2 cm in diameter, umbonate, hypogeal, with a prominent basal mycelial mat which falls away on drying to leave a marked scar. Exoperidium three layered: mycelial layer papery, pale amber, smooth to squamulose, peeling off in old specimens, debris encrusted in patches; fibrous layer leathery, thin, persistent; fleshy layer thick, buff to pale rust colour, on drying turns sepia, cracking, particularly across the base of the rays, the cracked edges tending to curve away from the fibrous layer. The fruit body opens by splitting along five to eight lines of weakness in the exoperidium thus forming the rays, eventually the rays are bent backwards exposing the endoperidium. Endoperidium sessile, up to 1·5 cm in diameter, globose, smooth, pale, vinaceous-grey to vinaceous-buff; peristome broad with fimbriate margin, silky, darker than endoperidium; Gleba amber; pseudocolumella ovoid; capillitium pale to medium brown, tapering at the ends, often encrusted; spores globose, 4–6 μm diameter, warted, pale to dark brown.

Distribution: cosmopolitan.

GEASTRUM PECTINATUM Persoon (Fig. 45.)

This is a very beautiful 'earth-star' and has been observed frequently growing on wet sandy soil, under the shade. The species can be easily recognised by its curled and long 'rays'. Fruit body 2–6 cm in diameter; exoperidium split into 5–9 cm

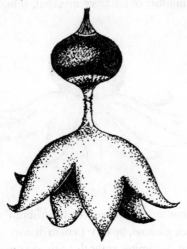

Fig. 45

long, pointed, unequal, recurved or spreading rays, usually bending backwards, raising the endoperidium to a height of 1–2 cm from ground level and breaking all its connections with the mycelium in the soil to which the rays are attached;

mycelial layer thin, pale brown, debris incrusted in patches, non-persistent; fibrous layer thin, papery, light-brown to whitish and persistent; fleshy layer fairly thick, non-persistent, nearly reddish-brown, darker when dried and very tough, endoperidium pedicellate, depressed-globose, 1·2–2 cm in diameter, smooth, brownish grey, darker brown on drying, occasionally tinted with blue, and sometimes looks as if it is dusted with powder; base of the peridium grey, forming a circle of radial ridges and grooves surrounding the pedicel; pedicel significant, distinct, 4–10 mm long, 0·8–1·5 mm wide, smooth, occasionally the remains of the fleshy layer form a ring round the pedicle; peristome silken, conical, fibrous, very distinct, 3–5 mm long and 2–5 mm wide, concolorous or paler than the rest of the endoperidium, surrounded by a groove, often with a basal apophysis; gleba powdery, pale-brown, capillitium 5–9 μm in diameter, encrusted with debris; spores 6–7·6 μ.

Distribution: cosmopolitan.

GEASTRUM DRUMMONDII Berkeley (Fig. 46.)

This species usually grows on the ground in dry places. The immature, unopened, sporocarp is globose and epigeal; exoperidium split in the middle and forms 8–12 'rays' which are subequal showing hygroscopic movement, expanded when wet, and folding over the endoperidium when dry; inner layer fleshy, amber, farinose, adnate, continuous; outer layer mycelial, thin, whitish, covered with

Fig. 46

debris, sometimes becoming detached; endoperidium sessile, or occasionally possesses a short pedicel, globose, up to 1·5 cm in diameter, dirty white in colour, finely asperulate, often becoming smooth with age; peristome boldly sulcate, often appear darker than the surrounding endoperidium; gleba ferruginous, pseudocolumella not evident; spores globose or subglobose, moderately and irregularly verrucose, often pedicellate, 4–6 μm in diameter.

Distribution: Africa, Australia and Tasmania.

GLOSSARY

Adaxial, describing the inner surface of the basidiospore next to the long axis of
   the basidium: usually that of the apiculus.

Aequihymeniferous, having hymenium development equal all over the surface of
   the gill.

Agglutinated, with hyphal wall stuck together.

Aguttate, spores and other cells not containing oil-drops.

Allantoid, sausage-shaped.

Alutaceous, the colour of buff leather.

Alveolate, marked with somewhat six-sided (honeycomb like) hollows.

Alveolus, small depression or hollow in a surface.

Amphigenous, making growth all round or on two sides; hymenium borne on
   all sides of the basidiocarp except the part which is directly attached to the
   substrata.

Amyloid, (see chemical tests).

Anastomosing, branching.

Angiosperm wood, wood with vessels, fibres, etc.

Annular volva, the lower outer part of the double annulus, infact it is a volva.

Annulate, having a ring.

Apiculus, the peg like projection of the basidiospore attaching it to the sterigma.

Apothecium, the cup or disc-shaped fertile part of a Discomycete.

Appressed, flattened onto the surface.

Arcuate, bent like a bow, arc-like.

Areolate, divided into small segments by cracking.

Ascocarp, general term for a fruit body producing asci.

Ascus, sac-like structure bearing spores internally.

Attenuate, narrowed.

Ballistospore, spore which is violently discharged from its mother-cell.

Basidium, the spore mother-cell of basidiomycetes, bearing spores on short
   sterigmata.

Bilateral, on two sides.

Binding hyphae, thick-walled, much branched; interwoven and often coralloid
   hyphae; these hyphae bind the generative and skeletal hyphae together.

Bitunicate, an ascus in which the inner wall is elastic and expands greatly beyond
   the outer wall at the time of spore liberation.

Caducous, falling away early.

Caespitose, growing in tufts.

Callote, smooth disc on the cap.

Campanulate, bell-shaped.

Capillitium, mass of sterile, thread-like hyphae mixed with the spores.

Capitate, having a head.

Cartilaginous, firm and tough but readily bent.

Caulocystidia, hyphal ends on the surface of the stem resembling cystidia or sterile basidia.

Cheilocystidium, cystidium on the edge of the gill.

Chlamydospore, thick-walled, asexual spore formed by enlargement of hyphal cell or cells.

Chlorohyaline, faintly greenish tinted.

Clamp, clamp-connection, characteristic protuberance at the cross-wall in certain hyphae of some basidiomycetes.

Clavarioid, having the appearance of *Clavaria*.

Clavate, club-shaped = clariform.

Claviform, narrowing towards the base.

Coalescence, fusion between two or more basidiocarps or hymenial areas.

Collariate, have a collar.

Columella, a sterile prolongation of the stipe within a sporangium or other fructification.

Conchate, like a bivalve shell.

Concolorous, same colour as.

Concrescent, structures, becoming joined as they grow together.

Confluent, coming or running together by irregular extension of hymenial areas.

Connate, joined together.

Context, the fibrous tissue which makes up the body of the fructification in basidiomycetes and ascomycetes.

Coriaceous, with a tough or leathery texture.

Cortex, a more or less thick outer covering.

Cottony, with fine hyphae or fibrils as strands of cotton.

Crenate, shallow and finely wavy.

Crenulate, edge delicately toothed with rounded teeth; longitudinally wrinkled.

Cristate, with small pointed, conical, or tooth-like projections.

Crust-like, spread over, and forming usually thin, closely adherent layer.

Cupulate, hymenium lining the interior of a cup-shaped basidiocarp or pileus, the outer surface of the cup being sterile.

Cuticle, the outer most layer of cap or stipe.

Cyanescent, turning blue.

Cystidium, a sterile, conical, prismatic or hair-like projection of a non-vascular hyphae in the hymenium or on other parts in Basidiomycetes.

Decurrent, gills, running down the stipe.
Decurved, bent down.
Dentate, toothed.
Denticulate, having small teeth.
Depressed, middle portion at a lower level than edge of the cap.
Detersile, removable, so that the surface becomes naked.
Dextrinoid, when spore mass on glass slide turns red-brown in Melzer's iodine.
Diaphragm, of Lycoperdaceae, a membrane separating gleba from subgleba and confluent with the endoperidium.
Dichotomous, divided into two branches.
Dimitic, with two systems of hyphae.
Disc, the central portion of the upper surface of a mushroom.
Dissepiment, partition between the pores.
Divaricate, very divergent.
Diverticulate, having pocket-like side branches.

Eccentric, away from the centre, between the centre and edge of the cap.
Echinulate, spores with acute spines.
Ectal excipulum, the outer layer of the apothecium.
Effused, more or less irregularly spread.
Elliptical, spores, parallel-sided and rounded at the ends.
Emarginate, gills, cut or notched at the point of attachment to the stipe.
Endoperidium, the inner layer of the peridium in Lycoperdales.
Epibasidium, the upper portion of the basidial apparatus of the Heterobasidiomycetidae.
Epicutis, the upper layer of the cuticle.
Epigean, above the ground.
Epiphytic, parasites and saprophytes growing on definite plant remains.
Epithecium, a layer of tissue on the surface of the hymenium of an apothecium formed by the union of the tips of the paraphyses over the asci.
Erumpent, bursting through the surface of the bark.
Evanescent, fleeting; having a brief exsistence.
Excipulum, the outer layer of the hypothecium.
Exoperidium, the outer layer of the peridium in Lycoperdales.

Farinose, like meal in form or smell.
Fibrillose, composed of longitudinal fibrils.

Fibrils, small fibres.

Filiform, thread-shaped.

Fimbriate, with tassel-like edge.

Ferruginous, rust-coloured.

Flabellate, fan-shaped.

Flocci, cotton-like tufts.

Floccose, more or less cottony; downy, wooly, flaky.

Frondose, deciduous or broad-leaved trees; shrubs.

Fructification, any fungal structure which contains or bears spores.

Fugacious, soon disappearing, e.g. the annulus.

Fuliginous, sooty in appearance.

Fulvous, a yellowish-brown shade, tawny.

Funiculus, a thin cord by means of which peridioles of some Nidulaiales are attached to the basidiocarp.

Furfuraceous, scruffy, or with small, soft scales.

Fuscous, dusky.

Fusiform, spores, spindle-shaped; tapering at both ends.

Fusoid, somewhat fusiform.

Gemma, a thick-walled cell similar to a chlamydospore.

Generative hyphae, thin-walled septate branching hyphae with or without clamps, in fruit bodies with other kinds of hyphae.

Glaucous, dull green to greyish-blue.

Gleba, spore mass, the inner fertile portion of the fruiting body of the Gasteromycetes.

Globose, spherical, or almost so.

Gloeocystidia, cystidia with dense, usually irregular and with highly refractile, hyaline or yellowish contents, thin or slightly thick-walled, often long, more or less deeply embedded in the tissue of the hymenium.

Gregarious, in groups.

Guttate, Guttulate, spores with one or many oil droplets.

Gymnocarpic, the fruit body developing without an enveloping tissue.

Gymnocarpous, fruit body with the hymenium exposed before the spores mature.

Gymnosperm wood, wood with tracheids, bordered pits, lacking vessels.

Heterogeneous, a basidiocarp composed of more than one kind of hyphae.

Hispid, with short stiff hairs.

Homogeneous, a basidiocarp composed of one kind of hyphae.

Homoiomerous, having the pileus and stipe components more or less intermixed.

Hyaline, colourless, transparent.

Hygrophanous, having a water-soaked appearance when wet.

Hymenium, the fertile layer of asci or basidia along with other associated structures.

Hymenophore, the fertile under surface of the cap.

Hypobasidium, lower part of a basidium after it have begun to produce epibasidia.

Inaequihymeniferous, an agaric fruit body, having the hymenium development unequal.

Incised, pileus margin, as if cut into.

Incrusted, hyphae, with matter excreted on the walls.

Inflated, hyphae, with enlarged cells.

Infundibuliform, funnel-shaped.

Inner veil, the membrane which covers the gills of a young mushroom.

Inoperculate, ascus, which opens by an irregular spical slit to release spores.

Intervenose, pileus, veins in the interspaces of gills.

Isabelline, a light buff-brown colour.

Lacerate, as if roughly cut or torn.

Lacinate, edge, as if shortly cut into thin segments.

Lacunose, surface layer having sunken gaps, e.g. stipe of a *Phallus*.

Lamellae, gills on which some Basidiomycetes produce their basidia.

Ligulate, strap-shaped.

Lineate, long and narrow.

Lobed, having rounded dimensions.

Locule, a cavity within a stroma.

Medulla, the part of the fruit body composed mainly, or entirely, of longitudinal hyphae.

Medullary excipulum, the inner portion of the apothecium.

Micaceous, covered with bright particles.

Micron (μ); a unit of measurement equal to 0·001 mm.

Monomitic, with one system of hyphae.

Monophyletic, of a single line of descent.

Mycorrhiza, an association between fungal hyphae and roots of higher plants.

Normal basidia, without carminophilous granulosity.

Obovoid, spores widest in the distal half (away from the point of attachment).

Ochraceous, colour of ochre; dead yellow or iron-rust colour.

Oleocystidia, cystidia with an oily resinous exudate.

Olivaceous, colour of olive; greenish-brown.

Operculum, a hinged cap on an ascus.

Ostiole, stoma, mouth.

Ovoid, spores, widest in the proximal half (near to the point of attachment).

Pallid, pale, deficient in colour.

Papillate, with small conical processes.

Paracapillitium, hyaline, collapsed and septate. Hyphae, resembling capillitial hyphae.

Pellucid-striate, having a somewhat transparent cap so that the gills are seen through it.

Peridial suture, membrane joining the outer peridium of clathroid fungi to the receptacle.

Peridiole, glebal mass surrounded by a hard, waxy wall of its own, contains the basidiospores, but acts as a propagating unit as a whole.

Peridium, outer wall of the fertile portion of the fruit body.

Peristome, area surrounding the stoma.

Peronate, sheathed.

Pileocystidium, cystidium on the surface of the pileus.

Pileus, cap of certain types of ascocarps and basidiocarps.

Pilose, covered with easily visible longish hairs.

Piriform, pear-shaped.

Placenta, tissue which nourishes the spore after its discharge from the basidium.

Pleurocystidium, cystidium on the sides of the gills.

Plicate, folded into pleats.

Pseudocolumella, a more or less densely woven, central mass of polverulent capillitium; the portion of stipe extended into the pileus.

Pubescent, set with rather long hairs.

Pulverulent, powdered; as if powdered over.

Pulvinate, cushion-like in form.

Punctate, having minute warts or depressions.

Punky, rotten.

Receptacle, the spongy part of a phalloid.

Rimulose, having small cracks.

Resupinate, fruit bodies lying flat on the substratum, with the hymenium facing outwards.

Reticulate, spores, with a network of ridges on the surface.

Revolute, caps, with edges roled up and back; backwardly curved.

Rhizoids, root-like structures.

Rhizomorph, a root-like mycelial strand; hyphae growing together in a tissue.

Rimose, where hyphae of cap become slightly separated radially, showing under-lying tissue; abundantly cracked.

Rough, spores with uneven surface and indistinct markings.

Rubescent, blushing; reddish.

Rufescent, turning reddish.

Rufus, a brownish-red colour.

Rugose, with ridges; rugulose, finely wrinkled.

Saccate, shaped like an open bag.

Sapid, savoury; agreeable to the taste.

Saprophyte, organism living on dead organic material.

Scabrous, rough and peeling.

Sclerotium, a hard resting body, resistant to unfavourable conditions; may remain dormant for a long period of time and germinate upon the return of favour-able conditions.

Serrate, edged with teeth.

Serrulate, delicately toothed.

Sessile, lacking a stalk.

Seta, thick-walled, often coloured, rigid cystidium.

Setose; or *setulose*, bearing setae.

Sinuate, waved edge of the gills that are notched or curved near tne stem.

Skeletal hyphae, thick-walled, mostly unbranched and aseptate hyphae in dimitic and trimitic system.

Spatulate, more or less like a flat spoon or spatula.

Spherocyst, spherical cells present in the trama.

Squamose, scaly.

Squarrose, rough and scaly.

Stellate, like a star in form.

Sterigma, a small hyphal projection which supports a spore.

Sterile base, rubber-sponge like tissue at base of some fruit bodies of Lycoperdales.

Stipe, stem; stalk of fungal fruit bodies.

Stipitate, with a distinct head or pileus and a stalk, free of the wood or bark.

Stratified, in more or less horizontal layers, especially the tubes of perennial polypore.

Striate, etched with fine lines, grooves, or ridges.

Strigose, with long stiff hairs.

Stoma, the apical pore through which the spores are discharged.

Stroma, compact mass of vegetative hyphae bearing fruit bodies.

Sub-, as a prefix, signifies slightly, almost or somewhat.

Sub–biseriate, consisting mainly of a double layer of chambers.

Subhymenium, the layer of interwoven hyphae between the hymenium and medulla or trama, giving rise to the basidia.

Subiculum, a mycelial felt covering the substrate and bearing the fruit bodies (c.f. stroma).

Subulate, slender and tapering to a point.

Sulcate, grooved; furrowed.

Tawny, colour of tanned leather.

Thickening hymenium, hymenium thickened by outgrowth of new basidia growing over the previous ones.

Tomentose, with shaggy matted hairs.

Tomentum, a covering of soft hairs.

Trama, the fungal tissue forming the pileus or bearing the hymenium; the layer of hyphae in the central part of a gill, a spine or the dissepiment.

Tramal plate, structure supporting the hymenium.

Trimitic, with three systems of hyphae.

Truncate, with the end flattened and appearing as if cut off.

Tuberculate, with small wart-like pimples.

Tunica, of the peridiole of Nidulariaceae, the outermost hyaline layer.

Turbinate, top-shaped.

Umbilicate, pit or depression at the centre.

Umbo, a central broad swelling.

Unilateral, on one side.

Uniseriate, single chambered; stipe-wall of a phalloid, consisting of a single layer of chambers.

Universal veil, of agarics and Gasteromycetes, a thin membrane covering certain types of young fruit bodies, upon expansion of the fruit body, the veil tears open and its remnants may be seen in the form of scales on the surface of the pileus and in the form of a volva.

Veins, swollen wrinkles on the sides of gills and on the under surface of a cap of a mushroom between the gills, often connected and forming cross partitions.

Velutinous, velvety.

Venose, looking like veins.

Ventricose, swollen in the middle.

Verrucose, having small rounded processes or warts.

Villose, covered with long soft hairs which are not matted.

Vinaceous, wine-red in colour.

Vinescent, changing to wine-red.
Viscid, moist and sticky, glutinous.
Volva, the cup-like basal remains of the universal veil after expansion of the fruit body.

Wart, a scale on the surface of the cap.

BIBLIOGRAPHY

Ahmad, S. (1952) *Gasteromycetes of W. Pakistan*. Lahore, Punjab University Press, Pakistan.

Beeli, M. (1927a) Contribution a l'étude de la flore mycolgique du Congo II. *Bull. Soc. Bot. Belg.* **59,** 109.

—— (1932) Fungi Goossenisiani IX. Genre Lepiota. *Bull. Soc. Bot. Belg.* **64,** 218.

—— (1936) Lepiota, *Fl. Icon. Champ. Congo.* **2,** 29–45.

—— (1938) Étude de la flore mycologique Africaine. Note sur des Basidiomycetes recoltes à Sierra Leone par F.C. Deighton. *Bull. Jard. Bot. Brux.* **15,** 25–54.

Christensen, C. M. (1955) *Common fleshy fungi*. Minneapolis.

Corner, E. J. H. (1950) *A monograph of Clavaria and allied genera*. Oxford.

Dennis. R. W. G. (1960) *British cup fungi and their allies*. The Ray Society, London.

Dissing, H. & Lange, M. (1961) The genus Geastrum in Denmark. *Bot. Tid.* **57,** 1–27.

—— —— (1962) Gasteromycetes of the Congo. *Bull. Jard. Bot. Etat. Brux.* **32,** 325–416.

—— —— (1963) Gasteromycetales I. *Flore Icon. Champ. Congo.* **12,** 215–232.

—— —— (1964) Gasteromycetales II. *Flore Icon. Champ. Congo.* **13,** 233–252.

Donk, M. A. (1960) The genuine names proposed in polyporaceae. *Personia,* **1,** 173–302.

Dring, D. M. (1964) Gasteromycetes of W. Tropical Africa. *Mycol. Pap. No.* 98.

Fidalgo, E. and Fidalgo, M. P. K. (1966) Polyporaceae from Trinidad and Tobago. *Mycologia* **58,** 862–904.

Gray, W. D. (1959) *The relation of fungi to human affairs*. Henry Holt & Co., New York.

Heim, R. (1945) Les agarics tropicaux à hymenium tubule. *Rev. de Mycol.* **10,** 55–58.

—— (1951) Les Termitomyces du Congo Belge recueillis par Madame M. Goossens-Fontana. *Bull. Jard. Bot. Brux.* **21,** 221.

—— (1952) Les Termitomyces du Cameroun et du Congo Français. *Mem. Soc. Helv. Sc. Nat.* **80** (**1**) 1–29.

—— (1955) Les Lactaires d' Afrique intertropicale (Congo Belge et Afrique Noire Francaise). *Bull. Jard. Bot. Brux.* **25,** 45–50.

Heinemann, P. (1951) Champignons récoltes au Congo par Madame M. Goossens-Fontana: V. Hygrophoraceae. *Bull. Jard. Bot. Brux.* **33,** 421–458.

—— (1954) Notes sur les Boletinae Africaines. *Bull. Jard. Bot. Brux.* **24,** 114.

—— (1956a) Champignons récoltes au Congo Belge par Madame M. Goossens-Fontana II. *Agaricus Fries s.s. Bull. Jard. Bot. Brux.* **26**, 1–127.

—— (1956b) Agaricus I. *Fl. Icon. Champ. Congo* **5**, 109.

—— (1963) Champignons récoltes au Congo par Madame M. Goossens-Fontana V. Hygrophoraceae. *Bull. Jard. Bot. Brux.* **33**, 433.

—— (1966) Hygrophoraceae, *Laccaria*, et Boletineae II. *Flore Icon. Champ. Congo* **15**, 289–290.

Ingold, C. T. (1965) *Spore liberation.* Oxford.

Korf, R. P. (1957) Nomenclature Notes. II. on *Bulgaria, Phaeobulgaria* and *Sarcosoma, Mycologia* **49**, 102.

Le Gal, M. (1953) *Les Discomycétes du Madagascar.* Paris.

Morse, E. E. (1933) a study of the genus podaris. *Hycologia*, **25**, 1–33.

Pegler, D. N. (1966) Tropical African Agaricales. *Persoonia* **4**, 73–124, figs. 1–146.

—— (1968) Studies on African Agaricales: I. *Kew Bull.* **21**, 499–533.

—— (1969) Studies on African Agaricales II. *Kew Bull.* **23**, 219–249.

—— Rayner, R. W. (1969) The agaric flora of Kenya. *Kew Bull.* **23**, 347–412.

Ramsbottom, J. (1963) *Mushrooms and Toadstools.* London.

Reid, D. A. (1965) *A monograph of the stipitate stereoid fungi.* Weinheim. J. Cramer.

Seaver, F. J. (1928–1951) *The North American cup-fungi (Operculates)*, 1928. *Supplement.* 1942. *(Inoperculates)*, 1951. New York.

Singer, R. (1962) *The Agaricales in modern taxonomy.* Weinheim, J. Cramer.

—— (1964a) *Marasmius* congolais recueillis per Mme. Goossens-Fontana et. dautres collecteurs Belges. *Bull. Jard. bot. Etat. Brux.* **34**, 317–388.

—— (1965) Marasmius. I. *Flore Icon. Champ. Congo Fasc.* **14**.

Smith, A. H. (1949) *Mushrooms in their natural habitats.* Vol. 1 text. Vol. 2, 231 stereochromes for Viewmaster. Oregon.

Zeller, S. M. (1949) Keys to orders, families and genera of the Gasteromycetes. *Mycologia*, **41**, 36–51.

Zoberi, M. H. (1963) Effect of temperature and humidity on ballistospore discharge. *Trans. Br. mycol. Soc.* **47**, 109–114.

SPECIES INDEX

SUBJECT INDEX